First published in Great Britain by VISION Paperbacks,
a division of Satin Publications Ltd., London

VISION Paperbacks
20 Queen Anne Street
London W1M 0AY
E-mail: sheenadewan@compuserve.com
website: http://www.visionpaperbacks.demon.co.uk

Originally published in the US by New Horizon Press

Cover Image: ©2000 Nickolai Globe
Layout: Justine Hounam
Printed and bound in Great Britain by Biddles Ltd.

© 1999 Janet Starr Hull, MS, CN
ISBN: 1-901250-45-8

SWEET POISON

How the World's Most Popular Artificial Sweetener is Harming Us

- My Story -

by Janet Starr Hull, MS, C.N.

PRAISE FOR SWEET POISON

"'Sweet Poison' is the remarkable story of a remarkable person. Janet determined that aspartame consumption was the probable cause of her affliction – and then proved it to all, including me, by its disappearance after avoiding this chemical. Unfortunately, most endocrinologists and journal editors refuse to believe or publish this phenomenon. Yet, 'anecdotes' such as this have paved the way for significant advances in medicine and science." – H.J. Roberts, MD, F.A.C.P., F.C.C.P

"Now, years after my first encounter with aspartame, a significant number of my colleagues are becoming aware of the multiple problems with the drug and are using this knowledge in their diagnosis and treatment. This book will bring awareness to the general public alerting them to the problems associated with using aspartame." – James B. Hays, MD, Federal Aviation Administration Designated Medical Examiner

"'Sweet Poison' reveals the truth a government agency (FDA) and giant artificial sweetener industry don't want you to know – aspartame is toxic. Read it and learn. Then take action!" – Mary Nash Stoddard, founder of Aspartame Consumer Safety Network

"I am very supportive of Jan s efforts, and believe that she has chosen to fight a very important fight." – Erik Millstone, Science Policy Research Unit, Sussex University, England

"'Sweet Poison' is a provocative book documenting the authors life-threatening illness due to aspartame poisoning, and her search for the answers she desperately needed to overcome her illness. This is essential reading for those who raise their health and wellness." – Dr. James R. Johnston, Director of American Health Science University and National Institute of Nutritional Education (Sole institution in the United States to award federally required Certification in Nutrition)

Dedication

To my sons

DISCLAIMER

This book is based on my experiences, and reflects my perceptions of the past, present and future. The personalities, events, actions and conversations portrayed within the story have been reconstructed from my memory, hospital records, letters, personal papers, and press accounts. Some names and events have been altered to protect the privacy of individuals. Events involving the characters happened as described; only minor details have been altered.

FOREWORD

We can trust our government to protect us from all enemies, foreign and domestic. Right? Well, guess again. You can trust your doctor to make an accurate diagnosis and then prescribe an appropriate remedy. Right? Well, almost. If your child or your spouse seems out of sorts, not up to par, has aches and pains, complains about fatigue, and a few other vague symptoms, he/she must be anaemic, be under stress, or has parasites. Maybe so, maybe not.

When I started practice, armed with the latest knowledge of drugs and how to use them, I did what I had learned in medical school. Pain? I treated with pain killers. Sick? A prescription for an antibiotic. Hyperactivity? Rx for Ritalin or an amphetamine. Allergy? A shot of cortisone. We had it all. And people knew that we had the answers and the prescription pad.

A few smart-aleck parents, however, wanted to know why the kid got sick in the first place. Answer: "Maybe he has bad genes from the father's side of the family." As time went on I began to realise that there were reasons for symptoms. These children did not have a Ritalin deficiency or a lack of penicillin. They were being poisoned by air, water and food, and their immune systems were failing. My wake-up call came when I realised that almost all hyper children (now labelled ADD) were low in magnesium, essential fatty acids and some B vitamins. They were often needlessly treated with antibiotics for virus infections. That changed their gut flora. We had to start all over. Just a brief review of standard medical literature indicated that our topsoil was disappearing and with the processing of food, nothing was left in our store-bought food but sugar, starch, flavours and trans-fats (the bad kind). Store-bought food leads to store-bought diseases. When I made sure that children got whole foods plus some vital minerals, vitamins, and essential fatty acids, many of their symptoms disappeared. They were not depressed. They were less sick. They would concentrate in school. I began to ask the right questions.

"Is your child a Jekyll and Hyde type?" Alternating between good and bad behaviour is a give-away clue for hypoglycaemia due to sugar ingestion or food sensitivities.

"Did your child have ear infections as an infant?" Ear infections mean cow milk has produced mucus in the ear tubes.

"Is your child restless and hyper-alert?" When the teacher complains about the child's inability to sit still in the classroom, it usually means that he is low in magnesium.

And so it goes. I was realising that for almost every symptom that the parents complained to me about their child, there was a nutritional component involved. Ben Feingold discovered that many unruly children had reactions when they ate foods containing salicylate. MSG was a problem to many. We began to question everything that was not organic, or pesticide free.

When aspartame came on the market, the incidence of reactions to food shot up. People were passing out, becoming depressed, having seizures, and developing symptoms of thyroid imbalance (Read about Jan in this book). Deaths have even been reported. When the FDA (US Federal Drug Administration) is asked about these reactions, they have two clever answers: (1) Research indicates that it is safe, and (2) Aspartame is considered food, so the FDA does not have to take note of these reactions.

The mixture of these potentially toxic substances in aspartame is about 160 times as sweet as plain sugar. What a concept. People who need their food and coffee to taste sweet would not get the unwanted calories. But look at what happens in the body. When this sweet taste hits the tongue, the message goes to the hypothalamus in the lower brain and states, "Incoming sugar!" This nerve centre sends a message to the pancreas to squirt out some insulin in preparation for the sugary calories. But no sugar arrives. The insulin lowers the blood sugar, and, in some people, it could drop to a point where the brain cannot function. Pilots pass out at the controls in the cockpit. Some folks will have seizures. An occasional child will become hyperactive. Some people will get pounding headaches. Many are depressed when they swallow this stuff. Mary Stoddard, founder of Aspartame

Consumer Safety Network, got eosinophilia myalgia. Janet got a diagnosable illness: Graves' disease.

How many doctors ask the question, "What are you eating and drinking?" Even if the patient asks if it could be aspartame, the doctor reassures the patient with, "Nah. You're depressed. Let's try some Prozac." If treating a wild and hyper child, the doctor might say, "He needs Ritalin."

Janet Hull has been through all these hassles. You will not have to do the same Even if you do not have any reportable symptoms, you will be amazed at how much better you feel and sleep after two weeks of being off the stuff. That may be the hardest trick: to eat without inadvertently getting some of it in your diet.

Janet has put up one more flag to tell us that we must be as organic as possible.

Dr. Lendon H. Smith

ACKNOWLEDGEMENTS

To all who are just beginning a healing journey – may this book help open doors.

To Chad – what would life be without you? Your help with the book (and everything!) is invaluable to me. I love you so much.

To Gwenn Zylla – the greatest reader in the universe. Only you know how to put the finishing touches on my work! I couldn't submit a page without you. You were always there to support me through this very long and tedious process. May we have many writing projects ahead of us.

To the Trophy Club volunteer fire-fighters, paramedics, and EMS team – the good energy each of you puts into the world by caring for others gives me hope for the future. I am proud to have been part of the team and will remember all of you forever.

To my Dad – you taught me to believe in myself and to never give up what I know is right. Thanks to you, I stayed with this project for over seven years. Thanks for believing in me. You always have.

AUTHOR'S NOTE

'Sweet Poison' has undergone many transformations. Initially, I was simply keeping a journal about a strange disease that appeared out of nowhere. No one knew the cause of my life-threatening case of Graves' disease, and I was concerned that the proposed drastic solutions might not be the right ones for me. Before I agreed to them, I wanted to know why I was sick. I thought writing down my experiences would help me decide my fate. Indeed, it has done just that!

When I discovered the chemical sweetener aspartame caused my illness, my personal journal shaped 'Sweet Poison'. As I became more involved in the aspartame issue, however, I uncovered hundreds of documents contradicting the NutraSweet Company's claim of product safety. I unearthed research showing marked danger to pregnant women, to unborn foetuses, and to children.

I was outraged to learn that as early as the 1960s, FDA and government officials were aware of this research and the challenges to the safety and long term effects of aspartame.

'Sweet Poison' transformed into a quest to help others with similar experiences conquer the deadly effects of chemical sweeteners.

I believe it's time to step back and re-evaluate the FDA approval process; to acknowledge the power behind advertising dollars and media sponsorship vs. responsible journalism; to monitor sincerity within the American Medical Association information networks; to criticise the quality of

contemporary nutrition offered at American medical schools (or lack of); to challenge the bullying behind the approval of other alternative sweeteners; to re-examine the safety of saccharin; and to expose the media's "information blackout" concerning topics such as aspartame safety.

The Aspartame Consumer Safety Network (ACSN), along with many journalists, pilots, lawyers, and independent researchers, has tried to reach the public with the truth behind aspartame. It has been a long struggle.

Thanks to my pioneering literary agents, Jeff and Deborah Herman, and my enterprising publisher in the US, Dr. Joan Dunphy, readers around the world now have the opportunity to know the truth about aspartame.

The lack of available information exposing the dangerous side effects of aspartame forced me to dig for hidden truths while I was recovering from a deadly disease. I not only stood firmly against my doctor's advice, but wearily breached the unknown and stood my ground against the threat of death. All I endured can be sourced to one thing, the lack of information about aspartame – the absence of truth.

If telling of my experience helps others, it will give meaning to the suffering I have endured.

PROLOGUE

Flames roar over my head. "Get down! Get down" Lieutenant Skinner yells. "It's going to flash over!" As I crawled along the narrow hallway of the burning trailer home, my left fire boot slid off.

Damn it! How can I fight a fire if my boot won't stay on my foot?

I can't see a thing! I can't wear my contact lenses because they'll melt in my eyes, and I can't wear my glasses because then the oxygen mask won't seal around my face. These things don't really matter. In a structure fire, everything is pitch black and smoky. All I need to see are the flames.

The fire singes my eyelashes.

Looking up blindly, I greet a massive sheet of orange flame roaring toward me. Its colour slides forward like an angry ghost. A tornado-like rumble vibrates my body. It's coming straight at me with a strong force. Suddenly, a blanket of intense heat envelopes me. Flames reach over my head and, boot or no boot, I am fighting this fire.

I'm not supposed to be doing this; I'm not supposed to be here. I'm supposed to be dead. And not by flames.

Two years ago, I was poisoned.

CONTENTS

A BAD HEADACHE

It was a day like any normal day. For me, anyway. I was home with my three toddlers doing my favourite activity, the "mom" thing: playing with and taking care of my children and nursing the baby. I was also doing my not so favourite things: changing three sets of diapers, picking up an endless number of toys and doing laundry. It was almost nap time for the boys and me. With the new-born riding side saddle on my left hip, I walked over to the refrigerator and grabbed a diet soda. Sean, my three-year-old, asked for a bottle of apple juice. We all sat down to watch thirty minutes of Sesame Street on television.

Then a blinding headache struck. Strange, I thought. I never get headaches. Oh well. I must be tired today, I rationalised.

The headache got worse. Sitting quietly on the sofa with the baby on my lap watching Bert and Ernie, the room began to spin. It spun faster and faster, like riding the merry-go-round at the Texas State Fair. It was making me sick to my stomach. I guessed I must be coming down with something. Then I began feeling nauseous. I was going to throw up if my head kept dancing pirouettes.

I let the boys stay up a big longer so I could remain s till a few minutes more. Maybe a sip of cola would settle my stomach. I reached for my diet cola. Another sip. I felt worse.

"Sean, honey," I said to my oldest with a pathetic tone in my voice. "Will you go over to the window and pull the blinds down for Mommy, please. Mommy has a headache. The sunlight is making it worse." By now, my eyes were throbbing in agony.

Little Sean proudly looked up at me, smiling. His new assignment made him feel so big. Honoured, he marched over to the window and

1

pulled the shade down with a force that almost yanked the linen roll off the brackets. I was too sick to care.

The baby started to cry, and Alex, my middle child, cried out for some milk. Sean was now running to every window pulling each shade down with vigilant drive. I felt worse.

"Sean," I barked. "Please stop pulling down the shades so hard. My scolding crushed his enthusiasm. He started to cry. Alex was now bellowing for milk, and Brian, my new-born, was so tired he was screaming. The sound shattered my aching head as if it were breaking glass.

"Mommy! Mommy!" Their cries rang in my ears.

"Boys, please!" I squealed. "Mommy is so sick. Help me stand up." Desperate to get to my bed, I depended on a three-year-old and a one-and-a-half-year-old to come to my aid. I put the baby in his play-pen praying he'd fall asleep quickly.

With gentle help from the boys, I worked my way down the hall. Hugging the walls to keep from falling down, I directed my feet to the bedroom. The house was now spinning round and round. It was becoming harder for me to swallow. This headache was like nothing I'd ever experienced before.

I dropped weakly onto my tidy bed. Circles continued to reel around my pounding head. The headache pierced the middle of my forehead, driving painful spikes deep into my eyes. "Oh, God!" I cried out. "What's going on?" I wished Chuck, my husband, would come home. I needed some help.

My body trembled and twitched. I started to sweat. The pain in my forehead persisted. "What is this?" I murmured, nauseous and agonised.

The boys slowly shuffled my way. They looked scared. Seeing me must have been frightening to them. "Mommy?" they modestly asked in unison. "What's wrong with you? Are you going to die?"

Unable to lift my head to comfort them, I replied, "Oh no, boys." Speaking made my head pound. "I'm not going to die." I wanted to chuckle at their innocence, but the pain was too intense. Any movement started me spinning like a top. The pain in the middle of my forehead pounded.

"Boys," I said, "go get Mommy a wet wash cloth from the bathroom. Sean's second assignment today. Neither Sean nor Alex was tall enough to reach the sink, but they toddled off to the bathroom full of delight with a challenge that would occupy them long enough to give me a bit of peace and quiet. I heard Sean, the inventor, and Alex, the domineering one, engineering plans on how to reach the sink. I had no strength to arbitrate.

Suddenly, the head pain began to subside. By the time the boys returned from their unsuccessful mission, I was able to shift from horizontal to a sitting position. Slowly, I dropped my legs over the side of the bed and stood on wobbly feet. I felt like I'd been hit by a truck. Creeping down the hallway, I made my way to the kitchen, stopping to peek at my sleeping baby. He was fine, thank goodness. The boys teetered close behind. I got them some milk and myself another diet cola. A cold one. I sat on the sofa and took a deep breath.

Damn. What just happened? I was confused but not yet worried.

A couple of days went by. I seemed fine. I guessed my headache was a freak occurrence. Other moms with three young children? must get headaches, too, I told myself.

The following week I got another headache. Same setting. Home with the children, watching Sesame Street on television, diet cola in my hand. Then it was spin city. Aside from the pain, I was getting annoyed. "I don't get headaches," I repeated to myself.

The next morning I headed for the grocery store. "I'd better shop fast in case I get another headache today," I said to the kids as if they'd really understand. They had no idea what I was talking about. Doing something simple like buying groceries is a big deal when you have three kids under three years of age in tow. Usually I tried to shop when Chuck was home so he could watch the boys, but he got home too late these days. I had to take all three offspring with me. I hoped the store was ready. As we drove there, I tried to talk to the older two about behaving once they were there. They grinned back, mischievously I thought, but maybe I was imagining that.

I found a parking space and after a couple minutes of fumbling with car seats, we were crossing the parking lot and heading towards the front door. Inside the store I lifted Sean into the shopping cart. I hoisted Alex kicking and squealing onto my back. He protested being strapped into the blue-framed backpack. Brian, unaware of what was going on, slept in a front pack laced across my chest, his limp legs swinging with every step I took. "Okay, boys," I professed with uncertain fortitude, "let's go get 'em."

Pushing my family cargo and the groceries we'd accumulated down aisle ten, I bumped into my next door neighbour, Edith. I adored Edith. She was the "cup of sugar" kind of neighbour. Like family.

Sweet Poison

"Howdy!" I said as I smiled with surprise.

Edith had a disturbed look on her face.

"What is it? What's wrong?" I inquired with concern.

"Jan," she said, "you're always so well put together. Have you had your pants on backwards all morning?"

"Well," I laughed, not too surprised that I didn't look my best, "I thought I had it together." I winked at her and we began talking of other things.

Suddenly, in a split second, a tiny dot appeared in my field of vision. I blinked, but it didn't go away. It was a bright pin-hole-sized light inside my eye. Minuscule at first, the pinpoint began to grow. It got brighter, too. Growing longer and more brilliant, it magnified into an electrifying jagged line blinding me in a matter of seconds. The line was blurry and fuzzy and annoying. I saw the shifting line whether my eyes were open or dosed. It got so large it overcame my vision, eventually disappearing behind my eye. When it manoeuvred behind my eye, a horrendous headache struck. Pain like never before. Even more intense than my other recent headaches.

I cried out.

Edith didn't know what was happening to me. I had no choice but to lean against the grocery shelves, riding out the pain.

"Edith," I cried out. "Help me. Something's wrong with my eyes. Take Alex out of the backpack, would you, please?"

Although she was confused, Edith did what I asked without question. Totally liberated now, Alex ran up and down the store aisle, laughing and having the time of his life. Sean wanted to join his brother and squirmed to get out of the cart. Edith had her two children with her. They wanted to start running around, too. The scene was quickly getting out of control.

Meanwhile, the pain in my head intensified, and I started to sweat. I feared I might have to vomit right between the cereal boxes and the pop-up toaster pastries. Edith stood motionless, unsure about what to do next.

Eventually the pain subsided. I wearily opened my eyes and stood up straight. With Edith's help, I checked out my purchases. She insisted on driving us home. Gratefully, I accepted. At home she unloaded my groceries and gave the kids a snack while I lay down.

"I wish Chuck would get home. I don't know what's wrong with me," I said when Edith came into the bedroom to see how I was. "I can't cope anymore. I wish these headaches would just go away. But they don't!

They come more often and get worse over time."

"They're migraines," Edith interjected.

I guessed she was right. But migraines? Me? How could I be developing migraine headaches all of a sudden?

I had no idea.

I figured perhaps my hectic schedule could be the cause, and I vowed to get better control over the demands in my life. I had to blame something for these recurring headaches. Too little rest. Three toddlers pulling on me all day long. A husband who worked too much. Whatever.

However, no matter how I rearranged the demands of my life in the next few months, nothing helped. I was experiencing at least one bad migraine every day. How could I cope with the boys' needs when I could barely stand up because of the pain?

Before those first couple of headaches, I never had a problem sleeping. But I was having trouble getting any rest now. And stress. Stress had always pushed me into action instead of getting me down. I was always happy-go-lucky, a person full of energy, always on the go. I loved doing things with my three boys. I enjoyed gardening, taking the children on walks, playing with the dog. I really liked my life. That was, until now I was changing, both physically and emotionally, and my life was changing. The changes were intensifying, too, and not for the better.

I was losing control of myself day by day. I started having problems with the kids and with my marriage.

Handling the boys had become less rewarding. When they bickered and fought over toys or were messy, I used to see the humour in it. Now I yelled. When they dropped food on the floor, I cried.

I could not seem to keep the house straightened or food prepared.

I continually asked my husband to spend more time at home. "You're never around to help," I protested. "I need more support with the boys. I'm so tired."

"Okay, Baby," he replied blankly, but his attention was on the ten o'clock news.

I had no energy. Because I couldn't get a good night's sleep, I was worn out during the day. If the boys didn't take a nap, I got very cross. "Go to sleep!" I screamed at them more than once. "Give Mommy a break. I'm worn out. I'm so tired." When they finally slept, I'd fall into my own bed for any rest I could steal.

My husband didn't understand what I was going through. Gone by

7:00am and not home until after 9:00pm every day, he seemed tired and on edge, too.

Why was I acting like this? Chuck worked hard as a self-employed contractor. I knew this. So why was his routine suddenly bothering me?

"My schedule never bothered you before," he said.

"It does now!" I snapped back.

Something was happening to me. I didn't know why I was changing, but something mysterious was taking control of my life.

BACKDROP

I grew up in Dallas, Texas and had an enchanting childhood. I have one sibling, a sister. We're both adopted and not related. Beth was adopted three years before me. But I always considered her my sister. Growing up, we met movie stars, producers and directors, rode in limousines, and ate at fancy restaurants. My father, Fred G. Hull, Jr., was one of three National Division Managers for Metro Goldwin Mayer Motion Pictures (MOM, Inc.). He had a glamorous job and always came home with captivating stories about the movies and the famous actors and actresses, making childhood magical.

Daddy told wonderful stories: watching Gene Kelly and Fred Astaire dance at the MGM studios; attending the world premier of Gone With the Wind in Atlanta, Georgia, in the 1930s with Clark Gable and Vivian Leigh; seeing Lucille Ball and Desi Arnaz on the set. Daddy watched the Doris Day film 'Please, Don't Eat the Daisies'. I was on the set when Robert Vaughn and David McCallum filmed the television series, 'The Man from U.N.C.L.E.'. I'll never forget the night Daddy brought me home an autographed picture of Elvis Presley.

I have forever admired my father. I still look up to him in so many ways. I am very much like my dad, even though I am not his biological daughter. In fact, I even look like my parents. I am small framed like Mom and have my dad's big blue eyes and extroverted personality. Many times growing up, strangers would comment on how much "you look like your dad." Ha! Daddy and I would glance at one another and share a silent snicker, never confessing the truth. We loved our little secret!

I was always a very holistic person. It was just my nature. As a little girl, I'd never tell my mom if I had a sore throat or if I wasn't feeling well until my fever gave me away. Then, I'd still resist any medications she made me take. I've always believed that the body is capable of healing itself if given the right tools. Tools such as plenty of water, vitamin C-rich fruits and juices, rest, and time.

I never had a weight problem, either. Enjoying abundant good health, I had no headaches, allergies, monthly cramping, PMS, or any serious illnesses or problems. In fact, other than the measles, I didn't even have routine childhood diseases.

I assumed I'd get married one day and have one or two children. But it was a pipe dream far in the future.

Yet the future has a way of descending upon us before we know it. I married a man from Iowa who I met on a blind date. A girlfriend talked me into flying to Des Moines with her. She was dating Chuck's brother. Chuck wound up moving to Texas to marry me. A tall, good-looking man, Chuck always wanted to live in Texas, primarily to escape the cold Iowa winters. Exercising his solid Mid-Western work ethic, Chuck started a very successful paint contracting business in Dallas after he graduated college. He had a good business, and I was fortunate to have been able to stay at home with my children while they were toddlers.

I never baby-sat or changed a diaper in my life before having my own kids. Sometimes I feel I wasn't properly prepared for the demands of motherhood. Yet I had three boys under three years of age! Sean, my oldest, was three years old when Brian, my youngest, was born. At one-and-a-half, Alex was sandwiched in between. There were times I was so busy I didn't know if I was coming or going.

The headaches started shortly after Brian was born. The worst ones struck right in the middle of my forehead. Annoying pin-points of light in my eyes showed up along with them. In addition, I was becoming cross and moody. My skin was drying out, and I was losing more of my hair than normal. I was really tired, too. Plus, I was putting on a few extra pounds. I guessed it was motherhood. Of course, my mind spun other possibilities but I pushed them away. I told myself to "keep on keepin' on" with my daily routine. What else could I do?

The boys and I stopped over at Mom and Dad's for lunch one day. I was going to tell my parents about my headaches. I never got to do it. Daddy was barbecuing his famous "Freddie Burgers," giant hamburgers he invented way back. Two oversized hamburger patties sealed like a pie, "Freddie Burgers" are stuffed with mustard, shaved blue cheese, red ripe tomatoes, and fragrant onions. He carefully placed the patties with pride on his barbecue grill and seared the mounds of meat to perfection.

Mom took Sean and Alex to the store to buy them a toy while Dad and I remained home to prepare lunch. I situated the baby in his carry cradle, freeing my hands to assist Dad.

Dad went into the utility room to get his burgers from the spare refrigerator. He returned with a strange expression on his face. Slamming the glass dish on the table next to the baby in the cradle, he began to moan aloud. He buckled over and groaned louder.

"What's wrong, Pop?" I asked with growing concern. Just then, he gasped for air, obviously in pain. Instinctively, I rushed to him before I could identify what was wrong. He collapsed in my arms. I don't understand how I did it, but I broke his fall by catching him in my arms and gently lowered his 176-pound body to the floor.

"Daddy! Daddy!" I cried out. "Oh my God! Daddy!"

He was thrashing back and forth on the floor like a fish on a hook. His moaning pierced my daze as I jumped up and raced to the telephone to call for help. Nervously, I punched the emergency telephone number and asked for an ambulance. Then I returned to Daddy, who had now lost consciousness. Without thought, I immediately began CPR.

Between compressions, Daddy rotated his head in my direction and, with closed eyes, took a terrible laboured breath. Afterward, his head relaxed and fell to the side. Then, to my horror, he stopped breathing completely. His face transposed from a pale white to a bruised blue and began to swell. He didn't look like the same man he had been ten minutes earlier.

Daddy died in my arms while my baby slept quietly in his carrier.

"Daddy! Don't die!" I begged. I continued CPR. How long I should keep it up, I didn't know. I laboured to force deep breaths into his mouth and drove straight-arm compressions to his chest, breathing again and again and pounding repeatedly. I continued until I was dizzy but I did not stop.

Totally exhausted, I continued on pure adrenaline. Between breaths I pleaded, "Daddy, don't die. Come on, Daddy. DON'T DIE!"

Then he flinched, a sign he was responding to my aid.

Forcing a feeble breath, he eventually opened his eyes, pleated his brow, and asked in confusion, "What happened? What's going on?"

"Daddy," I answered, exhausted, "I think you had a heart attack. An ambulance is on its way."

Sweet Poison

"I'm fine," he insisted. "You shouldn't have called an ambulance." I was afraid he would refuse help from the emergency squad when they arrived. I probably would have too if I were in his shoes. I realised, Now I know where I get my stubbornness. I also knew I'd lost him only minutes before but, for a fleeting moment, I doubted that it even happened. Then I remembered his final breath as if saying goodbye to me.

The paramedics arrived. They immediately removed the cardiac paddles from their bag preparing to use them and inserted an oxygen tube inside Daddy's nose.

At the same moment, Mom came home with the boys. I tried to remain calm, buffer the situation from causing turmoil for the boys and Mom. Daddy was transported to the local hospital where they installed a pacemaker inside his upper chest. Daddy has survived ever since on that one pacemaker.

Daddy's life changed. Mom's life changed. My life would never be the same. Little did I know that Daddy's brush with death might have prepared me for my own fatal encounter.

YEAR OF CRISES

W ith three active toddlers to care for and teaching aerobic classes six days a week, I received a call from the chairman of the Geography/Geology Department at the University of North Texas. "Would you be interested in teaching freshman geography classes and co-ordinating the respective laboratories?" Dr. Williams asked. Well, of course! I was thrilled. Joining a four-year university was a great opportunity. I gladly accepted the offer and faced an added ninety-mile commute to my already hectic schedule.

I had received my master's degree in environmental science when I was pregnant with Sean. And then, in January 1989, when Brian was nearly two years old, I was returning to work as an adjunct professor. "Maybe going back to work will keep my mind off these pesky headaches. Maybe they'll even go away," I rationalised at the time.

I tried to blend my new full-time work schedule with motherhood. I arranged my classes around limited day care and still spent plenty of quality time with the boys every day. I maintained a regular exercise routine to keep in shape. This was when my problems first began to escalate beyond the headaches. Because of my busy schedule I developed habits I never had before. I ate on the run. There was no time to take care of myself. At home with the kids, my diet was fairly simple. After I started working, my diet got sloppy. I gained a few more unwanted pounds. I was busy! So busy, I unknowingly destroyed thirty-five years of perfect health.

Then I made the worst mistake. I started drinking more diet sodas. I had started drinking them just after Brian's birth, but only sporadically. I knew better than to use artificial sweeteners. As an environmental scientist, I am aware that synthetic chemicals are not

11

meant to be eaten. But after Brian was born, I wanted to lose the weight as quickly as possible and then with my new job, I was always in a hurry. Whenever I left campus, I treated myself to a cold diet drink for my long commute home. I cast aside all my previous training. Even as a child, I instinctively knew what was good and bad, natural and unnatural. But I wanted to lose weight and certainly not gain anymore. I would pay dearly for this mistake.

The headaches didn't stop. And more bizarre symptoms began to appear.

Not only did I continue putting on extra pounds, but I began to retain water weight, which made me look swollen and puffy all the time. My aerobic buddies started teasing me about "gaining weight." In the fitness industry, weight gain is seen as a sign of laziness. "You're not working out enough," they would say. So, on the days I didn't teach aerobics, I began jogging.

I also tried eating less and less. What I did eat was mostly diet stuff. Logical, right? Exercise more, eat less, and eat low-fat or non-fat and sugar-free.

I filled my kitchen cabinets with boxes of food sated with preservatives, vacuum packages of low-fat, sugar-free snacks, and litres of artificially flavoured, sugar-free drinks. The refrigerator was stocked with fat-free, sugar-free yoghurt, low-fat processed cheese, the lowest fat-free margarine on the market, and more lifers of diet soda. The freezer was lined with boxes and bags of low-fat, sugar-free weight-watchful entrées, frozen veggies, and fat-free, sugar-free ice cream.

I was watching my weight by eating fat-free, sugar-free junk, or so I rationalised. In truth I had fallen prey to creative and deceptive advertising. Of course, at that point I didn't know it.

In fact I achieved none of my goals. Not only wasn't I thinner, but I became more nervous and irritable. "What's with you these days?" my husband asked. "You are really hard to live with lately. Why don't you go see somebody?"

"Oh, I don't know," I'd reply. "I'm just over tired with the kids and all. I'll be okay. Just give me a bit more help with the boys, and I'll be all right."

I have a lot of responsibilities on my shoulders, I justified to myself. I didn't want to fight with Chuck, so I didn't say out loud the resentments I felt: I can't be expected to handle all of life's demands in my usual optimistic manner, can I? So what if I'm a bit grumpier than normal. After all, I am gaining weight and I feel lousy, and I don't know why. That's enough to put anybody in a bad mood.

But my mood swings intensified. I was out of sorts all the time now. And I was becoming severely depressed. Boy, this weight gain is really getting to me, I thought.

My sleeping problems persisted into nightly insomnia. Before that first headache, I had always gone to bed early, slept like a log, and popped out of bed in the morning with a smile on my face. I was one of those "damned morning people." Now, I continuously had trouble sleeping. I couldn't fall asleep and, when I finally did, I'd wake up repeatedly throughout the night. I was having awful nightmares for the first time since I was a small child, too. I was not getting enough rest. I blamed it on the boys, on my workaholic husband, and my busy schedule And the cycle continued.

When I had an idle moment, which wasn't often, I assumed that my work schedule, exercising more and eating less were catching up with me, because I was always tired and feeling weak. How could that be? I lifted weights and did sit-ups every day. Working out regularly should have made my body stronger, not weaker. But I was weaker, without a doubt.

And my weight-gain didn't slow down. Even to myself I had to acknowledge my husband was right. I was hard to live with. I was living my worst nightmare!

I know humans shed hair seasonally just like any other mammal, especially in the spring and autumn, but I was soon pulling massive chunks of hair from my head every day.

Next, my fingernails started to split and tear. I'd always enjoyed long, hard, beautiful fingernails. Okay, I asked myself, what was going on: headaches, weight gain, hair loss, and now my nails? I was falling apart. It took a few more months before I realised these changes were simply not going to go away. All of them crept up on me one by one, so I didn't see them as parts of one problem.

The symptoms were irritating as hell, but I continued to shrug them off as by-products of the life I'd chosen. You name it, I blamed it. Excuses worked for a while, until one day my heart began to beat out of control.

For the first time in more than a decade of aerobic training, my resting heart rate uncharacteristically elevated. Not to mention that it skyrocketed during aerobic workouts. It actually hurt as my heart forced the blood through my veins. My irregular pulse made me feel dizzy and overheated. My aerobic workouts were becoming a strain, but I couldn't let my fitness students know this. In addition, I had an extremely difficult time maintaining my balance, and one day while teaching a low-impact class, I stumbled and fell. There I was leading a side-to-side grapevine with a packed class imitating my every move and boom-flat on my ass. I looked around but there appeared to be no uneven flooring or other reason. How embarrassing! I didn't know what to say so I laughed it off. But something happened to me then that had never happened in over a decade of aerobic training.

I perspired now more than I ever had. I literally had sweat streaming down me when I worked out-my leotard stuck to my wet body like plastic wrap. My students came to me after class and asked with sincere concern, "Are you all right?" "Oh, yeah," I replied as I try to think of some clever excuse for my awkward appearance. "I'm just going for it tonight."

My breathing wasn't normal anymore, either. I developed allergies for the first time in my life. I started using an asthma inhaler for what the doctor diagnosed as "exercise induced asthma." I could no longer complete a morning aerobic workout without taking a hit of medicine before and after class.

Questions ricocheted through my anxious mind. Could stress cause me to develop breathing problems? What else led to asthma for the first time in thirty-five years?

I didn't get much reaction from my husband. I wished he would help me figure this mystery out, but he was very non-committal. The boys were, of course, too little to help me. I was on my own with this one.

My physical appearance continued to deteriorate. I gained more weight and developed more puffiness. My eyeballs now protruded, causing difficulties with my vision. I'd worn hard contact lenses since I was fifteen years old, but other than that, my vision had been stable. I visited the eye doctor trying to find what could be wrong. "Why am I having problems now?" I questioned the eye specialist. He didn't know, but he did find some deterioration of the retina in both eyes.

My symptoms worsened. And so did my marriage. My health deteriorated from the inside-out; what I called the "silent kill." Slowly. Silently. The body deteriorates cell by cell, but you don't know what's happening because you can't see it. You feel bad, but nothing shows up in laboratory tests. Finally, over time, the symptoms manifest into a major disease you can see. A degenerative disease.

"I know something is very wrong," I voiced to myself. "But what is causing these problems?" I didn't realise these symptoms were all connected somehow. Headaches. Eye problems. Mood swings. Weight gain. Hair loss.

By now, my periods were so deranged I thought I was pregnant every other month, even though I'd had a tubal ligation a few months after Brian was born. Something was definitely awry, but I still blamed it on stress and my busy schedule. I frequently scheduled appointments with my gynaecologist. Every time I saw him with the same complaints of spotting throughout the month and bad cramping for the first time in my life, I was relieved that I was not pregnant. "So what's the deal, then?" I questioned. "I suspect I am developing endometriosis."

He disagreed. In fact, the doctor never found anything wrong with me. "You have all the textbook symptoms," he said, "but I can't find anything wrong."

"Okay, Doctor. I'll stop worrying about it and move on."

And that's just what I tried to do. I kept teaching at the university along with teaching aerobics and taking my morning jogs. The boys grew bigger and became more demanding. I kept gaining weight and continued having headaches. My husband still worked marathon hours and was home very little. The impasse between us deepened. And, every afternoon before I left the campus, I grabbed a diet drink for the ride home...

My symptoms kept mounting. I assumed things couldn't get much worse, but I was wrong. My emotions transformed from disagreeable to hysteria to my and my family's horror. "Have you gone to the doctor lately?" Chuck asked. "What does the doctor say?"

"Nothing," I replied in frustration. "No one can find anything wrong with me."

"Well, I wish somebody would," he said under his breath. But I heard him anyway, and silently agreed.

I knew I was hard to live with these days. I couldn't help it. Not only did I feel horrible and look like hell, but I rode an emotional roller coaster from the minute I woke up until the time I went to sleep.

One day while cashing a check at the bank, the teller asked me for my driver's license. I bit her head off. "Why do you want my license?" I felt my voice rise. "Do you think my check is going to bounce or something?" Mail she thinks I stole the cheque book. What's her problem? My temper flared without warning, and I embarrassed myself by becoming belligerent and acting the fool again. I immediately followed my verbal assault with an apology. "I'm so sorry.

Sweet Poison

I just don't know what has gotten into me these days. I guess it's my job or my kids. I don't know anymore."
And I was right. I didn't!
I was really distraught at this point because I knew I was not myself. I didn't understand what was causing my erratic behaviour or what would trigger me next.
Poor Sean, Alex, and Brian. They were so young and so sweet but despite my overwhelming love for them, I now found myself screaming and yelling at them for the most trivial things. Sean spilled his milk one night during dinner and I went ballistic. I couldn't seem to cope with the boys' normal needs and simple childishness. In the few hours he spent at home, Chuck seemed to be avoiding me. The reality of how bad things had gotten between us penetrated at times but I told myself, Maybe I'm imagining all this. Maybe he's not around enough to even notice. Maybe he doesn't care. I'm most afraid he really doesn't.

●●●

All these changes happened to me over one year. Just one year! First, the headaches. Then the weight gain. Next, my hair loss. Then my erratic mood swings and ongoing depression. And after that, my periods became irregular. One by one, my symptoms accumulated. Why had my life changed so drastically in just twelve months?
Why?
There seemed to be no answer.
I went to doctor after doctor. None could determine a physical cause for any of my problems; so I dismissed the seriousness and continued to blame my lifestyle for my failing physical and emotional condition. Life's stresses, my job, finances, my marriage, the kids. The same stale excuses.
One morning I woke to feel my heart skipping every sixth to seventh beat. "What the hell's happening?" I cried out. Chuck slept on. "This has gone too far. I have to get some answers."
I went to see a family doctor whom some other professors spoke highly of down the street from the university. After a thorough examination, Dr. Baker asked to perform an extensive thyroid scan. "Something's not right," he said almost too casually. "There's an increase in thyroid activity which may be the reason for your health problems."
"This never happened on any tests that were run by the various

physicians I've been to over the past year," I replied. "But I am relieved someone has given me some sort of answer for my recent health problems."

The doctor prescribed expensive thyroid medication and told me to take it easy. I went home feeling relieved. I'd made some headway.

But nothing changed in the next few weeks. I still experienced sudden migraines, PMS, continuous spotting, depression, and unpredictable mood swings. My hair kept falling out in huge clumps. My fingernails were down to nubs. My weight was now up thirty pounds. My skin looked irritated and was broken out, and my eyes protruded to the point that I didn't recognise myself in the mirror. Nonetheless, what scared me most was my heart. My heart now continuously skipped beats, sending unrestricted surges of blood through my veins.

Then, late one night, my heart abruptly began to beat ferociously until I lay motionless in a pool of sweat. I knew I had to do something else at this point. I thought I was about to die.

SWEET POISON

D amn it! I knocked the alarm clock off the bedside table as I feebly reached to see the time. It was 4:00am. Chuck and I had returned home from a weekend "getaway" only a few hours before – an attempt to rekindle something, anything of our relationship. Unfortunately, but not surprisingly, it wasn't exactly a huge success.

Suddenly I felt awful. What was happening to me now? I felt hot, very hot, and it was getting harder to breathe. As I lay in bed trying to sleep, my heart began pounding with such fury I actually saw my chest heave with each amplified heartbeat.

Sweat poured down my body, yet cold chills made my jaw chatter. "I'd better check my heart rate," I murmured. I reluctantly placed two lungers on my neck, afraid of what I'd find. I had reason for fear: my heart was beating one hundred eighty beats a minute! One hundred and eighty beats a minute? Propping myself up with a pillow, I cautiously counted again. One hundred and eighty beats. Now I was scared, really scared. I was slipping out of control. What was happening to me? What was wrong with my heart? Don't panic, I told myself. You must NOT panic!

Then I thought to myself, I'll just go to the hospital emergency room. They'll be to slow down my heart rate, and then I'll come right home. I could leave a note for Chuck, I thought. The boys would never know I was gone.

I rolled to my side and elevated my body up and out of bed with a weak push. I dressed, swished mouthwash around my mouth, and whispered to my still sleeping husband that I was driving myself to the hospital. I was not sure if he heard me or not. That was okay, really,

19

because I'd rather go through this alone, at least until I figured out what was happening to me. And if it was just paranoia, I'd only be embarrassing myself in front of some emergency room strangers.

I slipped out of the house and into the chilly morning.

The ten-mile drive to the hospital-seemed a hundred miles. It was drizzling. The roads were slick. I had a hard time seeing the road. I didn't see well at night, anyway. "Maybe I should just turn around," I murmured. "Yeah. 'Just turn around.' Jan, you're acting crazy."

At last, I pulled into the hospital parking lot and spotted a parking place near the emergency entrance. Feeling very weak, I struggled out of the car and walked timidly toward the door.

"I feel like I am dying," I whispered hoarsely to myself.

The glass doors slid open with a metallic bang, sending shock waves down the emergency room hallway. The smell of childhood vaccinations stung my face as I stepped inside and looked around. The room was almost deserted and creepy at this hour of the morning. I can't believe I just drove myself to the hospital. I hate hospitals! I think I'm going to throw up. I think I'm going to die.

I shuffled to the receptionist's desk. Sweat rolled down my back as I shivered with cold chills. A tired looking, middle-aged woman stood behind an oversized counter flipping pages on a clipboard. She looked up at me over the top of her heavy specs. "Yes?" she said in a monotone. "What can I do for you?"

I could barely hear her over the sound of my heart roaring in my ears. I explained my situation as if confessing a crime. "I really don't know what's happening to me," I said helplessly. "I just know I'm real sick, and I don't know why."

The tired receptionist ushered me to a cubicle where I exchanged my sweat pants and tee-shirt for a skimpy hospital gown, its flowered blue fabric faded from one too many washings. As if she were the hospital's headmistress, she instructed me to lie down and wait for the doctor, and left. Listlessly, I obeyed, and fell instant prey to other emergency room personnel who came in and out bearing thermometers, blood pressure cuffs, and plasma kits.

Pain struck. Pain, terrible pain ripping down my right leg. Every time I sensed that wicked piercing pain I hoped the nurses would pick up a jagged spike on the EKG. They didn't. For months, I had felt extreme biting pain wrap itself around my right knee and repeatedly shoot down my shin. The pains appeared out of nowhere, always without warning. Why didn't the nurses see what I felt?

The white sheets of the hospital bed felt crisp and cool against my skin. As much as I hated injections, I didn't mind when they punctured my vein and began the IV, and I was greatly relieved when the EKG became an appendage. I felt secure with it taped across my chest, as if my racing heart would return to normal now that professionals were monitoring me. I drifted in and out of consciousness. I wanted to sleep, just for a bit. Sleep... I tried to drift off. I was too restless, though.

An emergency room doctor on duty that night entered the cubicle and came to my side. He was very young, much younger than I. I'm not comfortable with doctors who remind me of my kids. I prefer the old grand-fatherly type. He talked to me about tests and other measures they wanted to take, but suddenly I was very tired, too tired to keep my eyes open, and I sank into oblivion.

The next thing I knew, I awoke in a private room, startled at my whereabouts. The room was dark and felt ominous. I didn't remember coming there from the emergency room. I must have drifted into a deep sleep. I felt as if I were surrounded by an opaque bubble, the sticky kind my kids popped all over their faces. I couldn't quite get outside of it, and I was not sure if I really wanted to.

Two days passed as if they were brief moments in time. I wished my husband and the boys were there. They didn't come to see me. It's best, I guessed. The boys were so little. I didn't want them to be scared by seeing me in a hospital bed so sick and swollen. Plus, Chuck was too busy, or so he said when he phoned and explained he was taking care of the boys in every spare minute. I was so alone, though. My anxious thoughts were my only company. As I lay in bed pondering my situation, I became aware of the television turned on in my room. I was not paying attention to it until an advertisement for a pain reliever caught my eye. Immediately following that advertisement, another ad selling constipation medication bellowed through the room. I reached for the remote and started flipping channels and counted more than twenty advertisements in less than five minutes hawking medications for headaches, aches and pains, monthly cramping, arthritis, constipation, diarrhoea, baldness, and tooth pain.

Then movie stars, professional ball players, and slinky models all bore witness to miracle prescription drugs recommended by doctors and available by "just calling your doctor for details." I never paid attention to commercials like these before. Until now, that is Lying in a hospital bed, deathly sick, changed not only one's outlook on life but one's habits.

A noise caught my attention. I spied a mysterious male figure lurking by my door, which was ajar. Maybe it was my husband, I thought, and propped myself upon one elbow. I was disappointed to see it was the ER doctor.

In the midst of pain, anxiety and sleepiness, I hadn't noticed much while I was in the emergency room, but now I was more aware of what was going on. The doctor was tall, fairly good-looking, and blond. He wore wire-rimmed glasses that were similar in style to the heavy horn rims the ER nurse had worn. With a sincere smile and a rehearsed manner, the doctor slid a chair next to my bed.

"You are a very sick woman," he said in his official doctor tone. "We need to have a little talk."

Uh oh, here it comes, I thought. His demeanour reminded me of the hospital protocol on daytime soap operas. I didn't know this doctor, yet I was being asked to place a tremendous amount of trust in him. Despite his youthful appearance, I did like him, though, and I felt safe in his care for the moment. I tried to concentrate, but was unprepared for what he had to say.

"You have what we call Graves' disease," he stated with textbook form. Startled, I sputtered, "What..."

"It's a disease of the thyroid gland, yet we don't know much about it," he said.

I thought it strange that a contemporary medical doctor didn't know the cause of such a disease. But I nodded as if this made sense.

He went on, "The thyroid is located at the base of the oesophagus in the throat and produces thyroid hormones." He spoke as if he were reading a script. "Thyroid hormones have a wide variety of effects on the body and are essential to life. They have many effects on metabolism, growth, and development."

My head throbbed as I tried to focus on the meaning of his textbook explanation.

"Graves' disease is also called thyrotoxicosis, a hyperactive thyroid gland that produces too much hormone," he continued to routinely explain. "Excess hormone production causes the body to remain overactive. All of the body's processes speed up, including digestion. This is why people with hyperactive thyroids typically lose a lot of weight."

His last words registered on me. "Lose a lot of weight?" I questioned. "I've gained thirty pounds."

Unaffected by this information, he continued, "The most common symptoms associated with hyperthyroidism or Graves' disease are nervousness, irritability, increased perspiration, insomnia, fatigue, weakness, hair and weight loss, separation of the fingernails, hand tremors, intolerance of heat, rapid heartbeat, and sometimes protruding eyeballs."

"Let's see. You're saying I have twelve out of thirteen symptoms. Why is it that the major symptom I don't have is weight loss? I've gained thirty pounds. My luck!"

He shrugged.

Headaches weren't on the list. I wondered why not. The mammoth headaches were my most distressing symptoms.

As the doctor leaned over my bed, I struggled to make sense of my situation. I knew something was very wrong, and just the name "Graves' disease" sounded so foreboding.

"I want to irradiate your thyroid and run some tests on your gall bladder," the doctor confidently continued.

My gall bladder? When did that come into the picture? I wanted only one problem at a time here! That was all I could handle, anyway.

"I must tell you that after we destroy your thyroid, I'll have to keep you on medication for the rest of your life to keep you alive," he added "But I can fine-tune you so you'll be better than before" He smiled a really big smile after he said this – almost sinisterly. I thought of Edgar Allan Poe's dark, malevolent characters as he rushed on. "You must do something about this soon; however, for you can die if we don't destroy that thyroid gland." He startled me as he abruptly pushed back his chair, its wooden legs scraping across the polished linoleum squares. Then he stood up as if, his job done, he was ready to leave the room.

'Wait a minute! Is this all you're going to say to me?" I cried out. "I have some questions. Don't run off! We are only getting started. First of all, what do you mean, destroy my thyroid gland?" I asked in desperation. "Is this the only alternative I have?" Who was this guy, anyway? Who was he to tell me what to do? Hey, I'd been through three natural childbirths, taught aerobics for over fifteen years, always eaten right (or so I thought at the time), didn't smoke or drink, blah, blah, blah. Didn't all that count for something? The doctor didn't seem interested in anything other than the standard diagnosis and my sketchy medical history he held on a chart in front of him. However, I wasn't as cavalier. Before he destroyed a vital part of my anatomy, shouldn't he study my daily routine to find a cause for this disease, I wondered.

Maybe not. The young doctor turned and silently walked toward the door.

I lay there stunned. Alone and confused, I had no clue as to why I was sick. The doctor didn't know. Nobody knew. I wanted to call my husband for advice, but he'd been so detached, why bother? He still hadn't come to the hospital to see me, and I couldn't tell my family because my dad was still recovering from his heart attack. I was being forced to make a decision about permanently destroying my thyroid gland with very little knowledge and no one to counsel me. Time out!

"Look," I yelled after the doctor, "I need more information and some time!"

From the doorway, he tried to quickly convince me that he could permanently solve all my problems by simply destroying my thyroid gland. Dr. Edgar Allan Poe hastily said, "Tomorrow I can have a specialist administer a finite dose of radioactive iodine that will 'kill' your thyroid." He tried to convince me that radioactive iodine was the best "thyroid assassin" because it was a simple and convenient treatment. I stared at him transfixed. In other words, it was a quick kill. Easy for him to say – it wasn't his thyroid!

"Most thyroid specialists recommend radioactive iodine for all their patients over twenty-five years of age with Graves' disease," he said as he continued his attempt to win me over. "Radioactive iodine is usually given in capsule form, which doctors prefer over surgically cutting open your throat. If you decide to do this, it will take several weeks to take full effect and, during this period, you'll have to be in total isolation because you'll be radioactive."

Instead of accepting what my doctor said, I wanted to question everything. He was leaving something out of the equation. I needed to find out what! "I doubt your recommendations are the only ones available," I blurted out loud. "There must be alternatives, even though I, of course, have none at the moment."

He looked stunned.

I thought the doctor believed he was sincerely doing his best to help me. I, however, preferred to discuss alternatives before I irreversibly destroyed one of my body parts. A necessary body part, I might add!

"I'll feel much better about your advice," I remarked, "if you'd be willing to first explore with me the cause of my Graves'."

He shrugged, and his apparent lack of interest in this aspect was an important turning point in my final decision not to follow his advice.

I cried out, "This is insane! I am not going to kill my thyroid with radioactive iodine! What happens to the poison once it leaves my thyroid? What else does it destroy on its way out? And don't tell me it won't do the rest of my body any harm!"

Taken aback by the force of my words, the doctor slowly inched his way farther out the door. He quickly added as he moved out of view, "You'd better do something about this soon, for you are in danger with a thyroid as overactive as yours. You cannot live with vital signs as high as yours are right now. Think about this for a couple of days if you have to, but I wouldn't take any longer. I can be ready to irradiate you in twenty-four hours."

When he finally disappeared, a multitude of feelings simultaneously rushed through me. "I can't suddenly come down with a deadly disease with no known cause or cure!" I cried out to the empty room like a defiant child. I didn't buy his explanation, but I was too exhausted to think about it anymore. I was being held captive in my unfamiliar bubble of confusion. I laid back and pulled the sterile hospital sheet up to my chin.

I wanted to sleep. I closed my eyes in hopes of sleeping forever, in hopes of escaping this nightmare.

It didn't take me long to fall into unconsciousness, and as I slipped away, I descended somewhere far, far inside myself, thinking, here I will search for answers.

I knew I had to find them and quickly, or I would die!

RISING FROM THE GRAVES'

And all the kings horses and all the kings men couldn't put Humpty
Dumpty together again.

I awoke after a few hours of hard sleep with images of Humpty
Dumpty running through my brain. I went over my symptoms for
the umpteenth time. Foreboding thoughts ricocheted through my
mind. Since no one else could or would, I decided that it was up to me
to put myself back together again. I definitely couldn't turn to my
husband for support – he was now avoiding both my problems and
me. My parents didn't even know I was in the hospital: I never told
them because of Dad's health problems. I wasn't sure how to do it, but
I couldn't resist the pun: I'll rise from the Graves' Gallows humour, I
grimaced, but any kind of laughter was better than none. My attempts
to find humour in my situation helped me to deal with the desperation
I didn't want to feel.

There was only one solution. I had to find the real cause of my
condition. I vowed not to make any permanent decision to destroy my
thyroid until I had answers. So I made my decision not to make a
decision.

In my mind I went over my symptoms one by one. One thing about my
condition that was jarring, especially since the doctor said weight loss
was a usual symptom of Graves' disease, was my illogical weight gain.
Was it some sort of due that something wasn't right about his
diagnosis?

I questioned the doctor several times about the inconsistency of my
weight gain in relation to Graves' disease. "Most patients who have a

thyroid as overactive as mine lose a lot of weight," I told him. "They don't put on thirty pounds."

His only comment: "Oh, you women! .Always worrying about your weight!" I wanted to blurt out what I thought of his patronising attitude. Instead I kept quiet. "You don't need to worry about what you look like right now. You need to concern yourself with getting well first."

"Well" to him meant losing my thyroid gland.

"But, if it's really Graves' disease, shouldn't I be losing weight?" I repeated in hopes of getting an answer. "Instead, I'm gaining weight. This makes no sense."

I never got him to focus on this clue that something other than Graves' disease could be causing my symptoms. He continued to insist my only course of action was to grant permission for them to destroy my thyroid. But for me it was a red flag.

Not only my self-image, but my life was changing without my permission, and I couldn't seem to stop it. I felt as if I had no control over myself anymore. Somehow, it had to stop, I told myself.

I made up my mind that I wouldn't do anything the doctor told me to do. Expecting to be judged as a defiant child, I informed him, "I am not going to irradiate my thyroid gland. Instead, I'm launching a campaign to find the cause of my Graves' before making any final decisions."

"You're making a big mistake," he said ominously.

"Perhaps. But it's my life, and I take responsibility for it."

Shaking his head, he left the room.

Though he had no interest in symptoms that didn't fit his diagnosis, he returned to quiz me at least a dozen times about my family medical history. Each time he asked, "Does anyone in your family have thyroid problems or diabetes? Have I already asked you this?"

My repeated reply, "I have no medical history, I am adopted," didn't seem to ever register with him. Though my questions to him appeared to get no reaction, he had aroused my curiosity about my absence of medical records and I thought I should try to contact my birth mother to ask her. However, the thought of doing that was a little scary and a little too much to think about at the present time.

Nevertheless, for the time being, I took a big chance defying his advice, elixirs, and pessimistic predictions. But I had to honour my own instincts. I knew that keeping my thyroid gland was the right thing to do. At least until I had more information.

For three days I had laid there with tubes and wires connecting me to

IV bags, EKGs, and sterile antibiotic drips. My immune system was so compromised by that point that I developed a serious upper respiratory infection. My blood pressure was too high, as well as my heart rate and cholesterol. Holding my lab results on a chart before him, the doctor asked with a puzzled look on his face, "What do you eat? You look so fit. Your blood levels don't match up, in my opinion." Equally as puzzled, I answered, "Well, a couple of days ago I had tofu for lunch. I always watch what I eat. Because I have been gaining weight this past year, I have been dieting regularly. I never had a weight problem until a year ago."

"Tofu? You eat tofu but have a cholesterol of three hundred?" he asked suspiciously.

"Yep," I responded. "I think I'm doing everything right. I watch what I eat, exercise every day, don't smoke or drink, never eat sweets or crave chocolate. I'm just as confused about this as you are." It was the damned weight thing again. I knew it was a major factor in this equation. I just couldn't figure out how to fit in.

"Why am I sick? What's going on here?" I asked. "Why have I gained so much weight when I am careful about my diet and exercise every day? Plus, I have an overactive thyroid gland! None of this makes sense. There has to be a reason for all this inconsistency. But what? I want some answers! And if you don't know," I said to the doctor, "I'll find out myself. I've had enough of lying around in this hospital."

Expecting an argument, I asked to go home.

He hit the roof. Well, sort of hit the roof for a well-trained doctor. "I disagree adamantly with your decision to go home." His face reddened. "You can't go home without doing something about your thyroid. A thyroid as overactive as yours is dangerous." He ran his eyes over the chart and wielded it like a stick. "Don't take this lightly, Jan." He stumbled over my name as if unsure of who I was. "I want you to really think about letting me irradiate your thyroid gland before you go home and possibly die."

"I can't," I answered, a bit perturbed that he wasn't sure of my name. "I just can't. Let me go home and I'll see you next week. I'll call if I get worse or something."

Reluctantly, he agreed. Immediately after he left the room, I gathered up my belongings, pulled on my sweats, and prepared to go home. "I hope my car is still in the lot," I thought with uncertainty. "I drove myself to the emergency room three days ago. I guess I'll drive myself home," I murmured.

Sweet Poison

Before I left the hospital I listened again to the doctor's warnings and instructions and filled the pile of prescriptions he loaded on me at the hospital pharmacy. I agreed to take all my medicine as directed until my final decision was made. If my thyroid lasted, that was. I also agreed to see him once a week for blood tests. His last words to me were a warning.

"Prolonged use of this thyroid medication could destroy your immune system. Yes," he told me, his eyes narrowing, "no one really knows the long-term effects of this medication. You need to make a decision about what you are going to do with your thyroid soon, because you could destroy your immune system in a matter of months. Then you'll really be in trouble."

I looked at it differently. The doctor had no idea what caused Graves' disease. He wasn't sure if the medication would kill me before my thyroid did, and if I did have my thyroid destroyed, he didn't know what the thyroid supplement would do to me on a lifetime basis. None of the options he was offering me seemed good. So I wasn't ready to grab at any of them.

I knew I was taking a chance walking out of the hospital, but I also knew that I needed time to find out if there were other options.

With a lurching gait, I stumbled down the hallway to check out at the floor desk. I waited for the nurse to process my final paperwork and, without really focusing, listened to her boiler-plate instructions on what to do when I got home. A few minutes later, feeling tired and weak I stumbled into the elevator and disappeared while she went looking for a wheelchair to push me out per hospital protocol.

Arriving home before Chuck picked the kids up from day-care, I examined my house as if I'd never been there before. Being diagnosed with a deadly disease abruptly changes the way one looks at life, I thought to myself.

Walking into the kitchen, I paused. The familiar smell felt like home. Tears formed in my eyes. Funny how scents can remind you of certain things: family gatherings, favourite meals, good and bad times. I was glad to be home, even though I was a bit unfocused. I tried to calm myself so I could appear normal for the sake of the boys. My husband barely noticed me anyway these days and didn't seem to care. But the children never saw me wired to tubes while I was lying sick in the hospital. I hoped they never would.

As I stood there galvanising what little strength I had left, I tried to think through my problem. "I guess I'd better find a way to strengthen my immune system while researching the cause of my Graves'," I declared out loud. "By becoming physically stronger, I'll encourage my body to defy the Graves' and the infections I keep getting. I refuse to surrender my health to anyone or anything!" I wanted to develop a plan to make the Graves' just "go away," but I was overwhelmed by trying to figure out where to begin.

I took a deep breath. The best place to start, I told myself, was with myself. I thought about my daily routine and began scrutinising what I did in a typical day as if looking for clues to a crime. "There has to have a reason why I have Graves' disease. I'll simply pick my life apart until I find it. And I will find the reason! I won't stop until I do." My heart skipped a beat and began to pound. I must find the answer fast, too, I realised. My thyroid might not last. I might not last.

I opened the pantry door and stood there staring at the packed shelves. I was not sure what I was looking for, but since I'd been steadily gaining weight over the past year, the pantry seemed like a good place to start. As an environmental engineer, I knew a lot about chemicals, so professional curiosity pointed me in this direction, too.

I eyed the boxed foods and began to scrutinise the labels sated with preservatives. I counted numerous vacuum packages of low-fat, sugar-free snacks and boxes of artificially flavoured, sugar-free drinks. Hmm. Then I wandered over to the refrigerator and peered inside. Let's see. There was fat-free, sugar-free yoghurt, low-fat processed cheese and margarine, bottles of diet soda, and packaged carrot sticks, which looked slimy. Opening the freezer, I studied the boxes and bags of low-fat, sugar-free diet entrées, frozen veggies saturated with fat-free cheese sauces to help them go down better, fat-free, sugarfree ice creams, and low-fat processed fish sticks.

My thoughts crawled in slow motion now as I tried to analyse what, if anything, was wrong with this picture. I shrugged. Nothing, I thought. I've simply been watching my weight by eating fat-free, sugar-free foods. But as I stood there, it dawned on me that I didn't really know what was inside these "wonder" packages. I began my new investigation by grabbing a piece of paper and a pen from the counter and assembling a list of all the chemicals in my foods. I realised for the first time in my life that proper nutrition in today's world is not what foods to eat, but what artificial, man-made chemical foods not to eat! I was no dummy, yet I had never been aware of the many chemicals, usually labelled as preservatives, found in just about everything. Maybe its the food chemicals making me sick, I thought.

Sweet Poison

Out of habit, I grabbed a diet soft drink and noticed NutraSweet's familiar swirl logo on the label. I looked at the ingredients and read the word aspartame. Without hesitating, I sipped the drink thirstily.

Before I finished the soda, however, I developed one of my migraines. That's when it came to me: the realisation that I only started to drink diet soda about a year or so before and that was when all my problems began. Another image came to my mind. A few days before I was hospitalised, I was driving home from the university sipping my usual cold diet drink which I bought every day before leaving campus, and a migraine hit. I had to pull off the road. I became excited. In fact, usually, before I finished the contents of a soda can, I developed a headache. Could the diet drinks be the source of these dreadful headaches? But how? Could this have anything to do with my Graves' disease?

Then another worry surfaced. If a diet soda could cause me such painful headaches, imagine how sensitive children's little bodies are to what's in a diet drink. Oh my gosh, I thought, I 've been feeding aspartame to my kids!

The more I thought about it, the more the words "fat-free" and "sugar-free" worried me. They were nothing but artificial additives flooding supermarkets. Perhaps fake foods saturated with chemicals did more harm to the body than good.

I walked over to the kitchen cabinet, grabbed another bottle of diet soda and studied the label. Why did I suffer a migraine immediately after drinking a diet drink? A coincidence? I didn't think so. A clue? It seemed possible.

How many chemicals did I eat in a day, I wondered? Even worse, how many chemicals was I unknowingly feeding my children?

I grabbed one of my sugar-free food packages from the pantry and read the label. Natural flavours, aspartame, di-sodium guanylate, partially hydrogenated vegetable oil, di-sodium EDTA added to preserve colour, TBXQ and citric acid in propylene glycol to help protect flavour, the bovine growth hormone, monosodium glutamate, and BHT added to help preserve freshness. I felt sick to my stomach.

Was this why I felt tired all the time, unpredictably moody, dangerously depressed, quick tempered, crampy, bloated, or fat? I didn't want to be sick anymore. I didn't ever want to see another squiggly line floating inside my head, exploding into a thunderous headache.

Could the answer be as simple as cutting sugar-free foods and drinks from my diet? Could this stop my symptoms? Banish my Graves' disease? I didn't know, but I had to find out. I prayed it was not too late.

DETECTIVE WORK

Back into the "mom" routine, I stopped with the boys at a friend's house a few afternoons later. Sitting on Deb's deck in a backyard full of trees always seemed to me like taking a vacation. Deb was a good friend from years past, and I always loved going to her house. So did the boys. They played happily taking turns in her hammock for hours. Over a cup of tea, Deb and I talked about my mysterious health problems.

"You know, I'm going to Doctor Steve Fugua, a renowned nutritionist in Dallas. My parents have been going to him for years, and they're very healthy and young looking." Deb told me. "Here." She handed me a piece of paper and a pen. "Write down his number and call him." Deb didn't have to twist my arm.

"I'll call Doctor Fugua tomorrow," I replied.

I didn't wait until the next day. As soon as I got home and put the boys down to nap, I picked up the phone and dialled his number. The phone rang. "Hello. Is Steve Fugua there, please?"

After a few seconds, a gentle voice answered, "Yes. This is Steve Fugua. What can I do for you?"

I introduced myself as a friend of Deb Crombie's and outlined my dilemma.

"I tell you what," Steve responded. "You come see me tomorrow morning at nine. I think I can help you."

Elated, I called Deb and asked if she'd watch the boys for me.

"Of course I will," she said. "Are you going to another doctor?"

"Nope," I answered with enthusiasm. "I hope I'm going to see someone who will help me once and for all." I didn't tell her it was the very person she'd recommended just in case it didn't work out.

Sweet Poison

But the minute I met Dr. Fugua I knew he was someone very special. He was tall with grey hair, a grey goatee, black horn-rimmed glasses, and a smile that immediately made me feel comfortable.

Steve Fugua was raised in his family's health food business, but his Ph.D. is in geology. Because of my master's degree in environmental science, we shared much. He taught me his unique knowledge of nutrition based on the influences of natural and synthetic chemicals. He drew his knowledge from over sixty years in the nutrition field. My knowledge of nutrition began with him.

After that first appointment, I saw Steve regularly. Sometimes the kids were with me, sometimes not. Every time I visited his shop, he seemed to be counselling someone. "Jan!" he often said as I walked through the door. "Come over here and tell these people about your Graves' disease. Jan is my best student," he would say to the people. Already feeling much better, I always smiled with warmth and admiration.

I had been avoiding anything containing aspartame since the day I'd returned home from the hospital. Under Steve's counselling, I decided to stay away from as many processed foods as possible, too, selecting instead whole foods at the grocery store for both snacks and meals. I replaced drinking all sodas with over fifteen glasses of bottled water a day. I became a food detective. I studied all food and drink labels. I learned what preservatives really are and how they are made. I became very protective of everything I put into my or my family's bodies, and modified their diets along with mine.

Sometimes when Chuck was home to watch the boys, I worked with Steve in the evenings. And I came to feel that other than my dad, there is no other male I admire as much as Steve Fugua. His ability to help people heal nutritionally is remarkable. I was fortunate that Steve took me under his wing and taught me the importance of nutrition.

Steve taught me more than any textbook could have about vitamin supplements, food additives, whole foods, and modern-day malnutrition. I asked question after question.

I began to understand how people pollute their bodies in the same way they pollute the environment. Chemicals aren't meant to be eaten and can accumulate in the body like toxic waste accumulates in a river. Hidden chemical food additives dominate the food supply, and fat-free, sugar-free food substitutes tempt us to forget basic eating rules. Instinctively, our bodies require real food – not food substitutes. I learned that I had been saturating myself with unhealthy chemicals. Steve spent hours not only with me but with each of his clients. He taught us about nutrition and shared his knowledge openly. Coincidentally, he had two other clients diagnosed with Graves' disease, both of whom he followed to complete recovery. I was number three. As our working relationship and my knowledge expanded, I supplied Steve with enough details to document patterned similarities among the three of us: two females and one male.

Every day I grew more conscious of what foods my body needed to maintain perfect health and which disguised artificial foods to avoid. I realised that my eating habits should be more like my grandparents' diet one hundred years ago.

Eat real food: that's what the body demands. I realised how far from common sense I had wandered.

Steve kept a journal of all his nutritional case histories. He had documented over sixty years of cases he still referred to. One day as I arrived at his shop, Steve grabbed my arm and whisked me out the door. "What are you doing?" I asked with a smile. "Steve, where are we going?"

"You'll see," he said with a sheepish grin. We got into his car and drove a short distance to the print shop. Unaware of what he was up to, I jumped out of the car and followed him inside. His stride reminded me of a teenager's. We walked over to the counter. Steve presented the clerk with a box filled with papers neatly stacked, though each page was slightly dog-eared and worn. He requested his entire journal be photocopied for me. I was speechless. He was giving me his life's work, the greatest gift he could ever impart. My throat quivered as I held back tears of gratitude.

"Steve," I said with a hug, "I am overwhelmed. Thank you so much." He smiled.

For decades, Steve had helped clients rid themselves of all kinds of ailments. Now he added my story to the list. As we stood there, he handed me my own case study. I began reading:

"But Graves' Disease Can't Be Cured"

"Jan is a very active thirty-four-year-old mother of three young boys. She has been perfectly healthy all her life, doing what she feels is right by eating little saturated fat, no butter, no eggs, few sweets, and consuming little alcohol. She teaches aerobics six to ten times a week, instructs environmental science classes at the local university, and maintains her household after working hours. She has fallen into the typical "fitness mode" of working out too much and eating too little in order to maintain a lean, strong body. She consumes a lot of diet

Sweet Poison

products sweetened with the artificial sweetener NutraSweet. Due to her busy schedule, she eats sporadic meals at irregular hours, eating too many 'low calorie' hydrogenated foods."

It was very strange seeing it all in print but I wanted to know all Steve had concluded and read on, intrigued.

"For almost a year, Jan's weight has slowly increased pound by pound. She keeps working out and eating less as her weight continues to climb. Her heart has begun to race, she sweats a lot, and her menstrual cycles are irregular. Her vision is worsening, she has retinal tearing in both eyes, and her skin and hair are drying out. As her body changes, she eats more diet and processed foods. She blames her mood swings (PMS all month long) on stress.

"Finally Jan ends up in the hospital with a racing heart rate of 180 beats per minute. She is stricken with a serious upper respiratory infection and hyperthyroidism. Her doctor diagnoses her with Graves' disease.

"Jan is told that she will die if she does not get her thyroid under control, and it is recommended to her to drink radioactive iodine (the radioactive cocktail) to destroy her thyroid gland. She then will be put on artificial thyroid medication for the rest of her life to keep her alive.

"Jan refuses to let her doctor permanency destroy such an important part of her body. She wants to fight to keep her thyroid. She checks herself out of the hospital after three days on an IV and goes home.

"Upon hearing from a friend about the successes of nutritional counselling, Jan comes to me with her prognosis of Graves' disease. She has a hair analysis performed. All of Jan's mineral and nutrient levels are very low, most likely due to the chemicals saturating her body, and she shows a lack of stomach acid available to dissolve them. Her chromium, zinc, selenium, manganese, magnesium, B-6, and vitamin C levels test dangerously low. Two other clients have been previously diagnosed by their doctors as having Graves' disease. Interestingly, all three are deficient in the same minerals and nutrients: vitamin C, PABA, selenium, zinc, B-6 and chromium.

"Once Jan brings these mineral levels back to normal, her Graves' disease disappears. Within thirty days, her heart and thyroid return to normal. Jan must maintain a high acid level in her stomach, however, in order to digest her food, her vitamins, and her nutrient intake. After more than thirty years of eating alkaline foods such as white flour and margarine products, Jan's acidic level within her stomach is too alkaline. After meals, she now takes a digestive enzyme rich in papaya or eats a raw lemon, drinks a rich red wine, or takes a betaine hydrochloride supplement to aid in the digestion of her food, especially if she chooses to eat red meat or a meal heavy in fake oils. To support the stomach lining, she eats raw cabbage at least three times a week to keep the mucin cells lining the stomach walls click and healthy. Research supports the theory chat stomach ulcers disappear when raw cabbage is eaten to rebuild the stomach lining.

"Jan eats 75% raw food with each meal, including high fibre grains, eats little or no red meat, drink an abundance of water, and maintains a regular supplementary program of the following vitamins and minerals: chromium (picolinate and glucose tolerance formula [GTF chromium]), zinc picolinate, PABA, pantothenic acid, vitamin C, liver tablets, selenium, manganese, calcium-magnesium, primrose oil, B-complex and extra B-6, along with a natural multivitamin."

I hugged Steve in appreciation of all he'd done for me and made a promise to myself that if I ever had the chance to help people who were ingesting sweet poison to regain their health and attain a healthy lifestyle, I would.

Each week as I had promised, I got a blood test at the ER doctor's office. We looked at the tests differently. I gauged the results of my personal efforts not to irradiate my thyroid gland at the same time as the doctor continued to make plans to irradiate my thyroid.

At my weekly examination one month after leaving the hospital, my astonished doctor announced somewhat sheepishly, "Your thyroid levels have returned to normal."

"Doctor," I said as he shook his head, "what do you think? Am I getting rid of this Graves' disease?"

"Well," he sputtered, "it's too early to tell. We might not have to destroy your thyroid," he said grudgingly. "But let me warn you," he went on as if looking for something wrong, "your thyroid levels will not remain normal. Absolutely not! No one has ever cured Graves' disease," he proclaimed with authority.

I refrained from comment. We'll see, I thought.

As my body continued to stabilise, I graduated to having blood tests every two weeks instead of once a week. I felt stronger physically and emotionally. "I am defeating this disease," I told myself hopefully, but I still had some concerns that the doctor's prognosis might come true.

After a few more weeks, just when I was becoming comfortably optimistic, something did go wrong. I developed uncharacteristically dangerous reactions, worse than before my hospitalisation.

Sweet Poison

My skin began to break out in grotesque acne. My hair fell out in bigger and more frightening dumps. My eyes weakened to the point that I had difficulty focusing on anything, and my night vision was completely elusive. I looked and felt worse than I did when I first got ill.

I was horrified but I tried not to panic.

Breathe deeply, Jan, you have to gain control. Time to rethink. Are the vitamins tricking me or is the medicine impairing my immune system? I went over all I was doing to try to find the culprit. Could taking both the medication and the vitamin supplements be overloading my system? I wondered. I knew my doctor would blame the vitamins. I could hear his reprimands: How dare you go against this advice? That's what you yet! I would be penalised for displaying self-assertiveness, and I would become even more sick. That'll show me, I thought.

Nevertheless, it could be the medication, not the vitamins. After all, I was warned the medication was so strong it could destroy my immune system.

And so I decided to take another bold step. I would stop taking my medication. I eased off it slowly, cutting back my daily dosage.

I established a target goal of three months to stabilise my condition, relying solely on good nutrition and vitamin therapy for recovery. If at that point my condition was still this bad, I would have to reconsider the doctor's treatment. I continued to supplement my diet with the vitamins in order to restore the damage caused, I was sure, by the aspartame. But I took no prescription drug, not even aspirin.

Without doubt, I was taking a chance following my instincts. I do not recommend going against a doctor's advice, but, in my case, I felt I knew something he didn't.

If I regained good health and maintained it for the next few months, I would know the dietary cleansing and vitamin therapy had healed my body.

TAKING CHANCES

I called my sister to let her know what was happening. When I told her about my Graves' disease, she astonished me by confessing she also had the disease.

"You never told me you had Graves' disease, Beth," I said. Beth had left Dallas after she graduated from high school more than twenty years before. Our communication was not always the best since she moved away, but I was embarrassed I didn't even know my own sister was diagnosed with Graves' disease. She began relating her symptoms. Beth had a form of Graves' disease that affects the eyes. Coincidentally at this time, President and Mrs. Bush and their dog were each diagnosed with Graves' disease.

The media rumbled about how unusual it was that non-blood relations within the Bush household had simultaneous cases of Graves' disease. The chances of this happening were around one in ten thousand, they reported. They also commented that thyroid problems are unusual in men and even more unusual in dogs! Men and dogs don't have the same endocrine system as women, making it uncommon for them to contract thyroid disorders like Graves' disease.

The trail of Graves' sufferers I knew or heard about was growing. I found it most curious that my adopted sister and I both had Graves' and that President and Mrs. Bush both had Graves'. Feeling like a detective I tried to follow the clues. Could the Bushes' dog have been exposed to the same trigger that caused his owners' disease? Table scraps, maybe? A taste of dessert, bread with jam? I knew the Bushes, especially Mrs. Bush, were very diet conscious. Did they use products with aspartame? It seemed likely. After all, didn't most people who had a craving for something sweet and didn't want to overindulge in that

41

culprit-sugar-choose a substitute like aspartame?

To date, no one has discovered a cause for Graves' disease, but at the time of the Bushes' thyroid problems investigators suspected faulty piping in the vice-presidential home might have contaminated their drinking water, possibly overloading their thyroids with toxic metals. Investigators were on the right track in tracing the cause of the Bushes' Graves' to an environmental overload of chemicals deposited in the thyroid, but they stopped their investigation short of finding a specific chemical.

Like them, I was becoming more and more convinced that my Graves' was caused by an overload of a toxic chemical. The question which they'd given up on was the same question I continued to ask. Which chemical? I wondered what chemicals were in aspartame and asked myself how I could find out.

Was I looking for a needle in a haystack? Luckily, I was competent in investigations like this because of my environmental background. As a professional engineer, it was my job to investigate toxic chemicals deposited in soil and groundwater. Could I pioneer an investigation to successfully expose the chemical or chemicals deposited in my body? I focused my investigation on the Bushes' two cases of Graves' disease, trying to trace any environmental similarities between the illnesses of President and Mrs. Bush and that of my adopted sister and myself I read every obtainable article and watched every interview with the Bushes in which they discussed their cases of Graves' disease, what the doctors did before and after they destroyed their thyroids with radioactive cocktails, what they ate, how they lived, where they travelled, their exercise routines, their personal histories, everything I could find for dues. I presumed that unless President and Mrs. Bush were travelling in separate cities, they most likely ate, drank, and slept in the same places.

How did this relate to my sister and me? Neither of us did the same things or travelled the same places as President and Mrs. Bush. So what did the four of us living in different surroundings with very different lifestyles do which was identical?

I considered whether my sister and my Graves' might source from the way we were raised. But I quickly discarded this idea. Mom brought us up on southern home-cooked meals and flu shots every year. Childhood dinners and midnight refrigerator raids had departed from our systems by now. Beth and I both agreed that the cause had to be something we did, and were still doing, after leaving home.

Maybe Beth and I, having been raised in the same household, now ate the same type of foods. I wondered if my sister's pantry looked like mine. We had lived in different cities for many years, but we still might eat similar things. We might even be eating or drinking the same things the Bushes did. We all might be causing the Graves' disease ourselves. Like "sisters in crime," Beth and I began scanning our diets. We compared foods that we both ate regularly. Nothing too unusual popped up, until I mentioned drinking a lot of diet drinks. Beth interjected, "That's funny. Lately, I've been buying a lot of sugar-free foods with the NutraSweet label. I'm always so thirsty. I sip diet drinks all day long because I'm developing problems with my blood sugar."

"I've been drinking them because of my weight problem."

Despite our motivations being different, to me, the fact that both Beth and I had developed Graves' after we both started heavily drinking diet sodas was more than coincidental. Within a few weeks of our conversation, Beth's condition worsened. She became an insulin dependent diabetic. One morning just after waking, Beth checked her blood sugar to find it so low she was rushed to the hospital, by that time nearly in a coma. Even though we were not blood sisters, I had no doubt I might soon follow in her footsteps. After all, I had duplicated her path of illness so far. That scared me. Again, what were the odds of adopted sisters both developing Graves' disease and both possibly diabetes? What were we doing the same?

I had never used anything with aspartame until a year or so before my Graves', but Beth confessed she had been using saccharin in her coffee and iced tea, until, after NutraSweet was introduced, when saccharin became harder to find. It was also around this time that saccharin started getting bad publicity, which worried her. As we talked, Beth remembered first developing blood sugar problems around this time. Now she was insulin dependent. Coincidence? I thought it had to be the diet sweetener. But how?

The year before my Graves' diagnosis, I fell into the trap of believing "sugar-free is responsibility-free." Advertisements convinced me I could eat and drink all I wanted without paying the price of gaining weight. "How could I have been so gullible?" Beth bleakly added, "But what else can a diabetic use?"

"We'll find something safe," I promised.

Meanwhile, my own body continued to stabilise. The purified diet, the vitamins, and no aspartame seemed to be returning me to health. I was more convinced than ever that it was the aspartame which had been making me sick during the past year. Now, I had to prove it.

I also was beginning to feel I never actually had Graves' disease. Maybe the real Graves' can't be cured. Perhaps my symptoms simply mirrored those of the disease. Could they be, instead, a reaction to aspartame's toxins accumulating in my thyroid gland that mimicked textbook Graves'?

As I learned more and became more convinced of aspartame's dangers, I also began hearing stories of others who had reacted negatively. I decided to write to the NutraSweet Company to tell them of my experience. When they wrote back they labelled my experience as anecdotal, which meant they didn't think it proved a thing because I was not an "official" laboratory guinea pig. I disagreed. I knew what I experienced, and my experience was real and could be summed up very simply: no more aspartame – no more Graves' disease. That was proof enough for me.

Moreover, knowing that there had to be too many other people similarly suffering as Beth and I had, made me want to do something, anything, to warn them of the dangers of aspartame.

I thumbed through the phone book and called a number listed under Speakers Organisations. If I was going to reach the public, I wanted to learn the best way. They recommended I talk to a woman who could help me get started, Mary Nash Stoddard.

Once again, fate seemed to direct me. A publicist by profession, Mary, I soon found out, had founded the Aspartame Consumer Safety Network (ACSN, Inc.) in 1987. It was a coincidence that astounded me. The ACSN is an international consumer safety organisation founded to inform the public about the background and questionable safety of aspartame (NutraSweet). Mary exclusively funds the organisation and dedicates countless hours day and night answering a toll free aspartame hotline.

When we met, Mary shared some impressive facts about aspartame which outlined the dangerous side effects the diet sweetener had on laboratory monkeys, rats, guinea pigs, and humans. She apologised for being so straightforward. After all, she didn't know me and we didn't meet to discuss the dangers of aspartame but to launch my speaking career. She felt this information might help me to help others who were suffering similar symptoms.

Mary quickly exposed me to the politics of the food industry, which was a real eye-opener. "I can't believe how naive I've been. Probably," I said, "millions of other people around the world like me believe anything sold for consumption is safe. Too many untrustworthy decisions directly affecting our health are being made by the wrong people: bureaucrats, politicians, big business executives. Not only was I being bamboozled, but millions of other people are being victimised." "There are some documents I need to show you," Mary said soberly. The documents she gave me were shocking. They showed that aspartame is found in thousands of food products. Research studies as early as the 1970s showed holes in the brains of laboratory mice fed aspartame! The FDA had received several thousand official complaints against aspartame. Three Senate hearings had been held to debate the public safety of aspartame, but nothing had ever been adjudicated.

The true history of aspartame was deeply disturbing. I was disappointed in the American government, notably the FDA, which I had always believed ensured the safety of the food we ate. However, I now learned that the FDA possessed the results of research which proved beyond a reasonable doubt that aspartame consumption was risky.

Among the documents Mary showed me was a letter written by Linda Tollefson in 1987 about aspartame complaints and safety, and printed by the Office of Nutrition and Food Services for Safety and Applied Nutrition (CFSAN) of the United States Food and Drug Administration. According to Tollefson, over 3700 complaints were received by CFSAN, of which, seventy-seven percent were from females between twenty and fifty-nine years of age. Of the seventy different symptoms reported, ten percent were considered severe, including headaches, seizures, and allergic, gastrointestinal and behavioural reactions. Tollefson offered two tables that broke down the reported complaints by the product they were associated with and by the symptoms that accompanied the reactions. Over 1900 complaints were associated with diet soft drinks - the most of any product reported. Other products reported to have adverse symptoms associated with them were: table top sweetener (1002 complainants), puddings/gelatine (348), lemonade (283), powdered fruit drink mix (268), hot chocolate (229) and iced tea (218).

Headaches topped the list of symptoms reported with over 1000 complainants. It was followed by dizziness/balance problems (461), mood swings (399), vomiting and nausea (364), abdominal pain and cramps (268), seizures and convulsions (211), and changes in vision (186). Since CFSAN gathered the data from victims and their physicians and not from a controlled laboratory setting, Tollefson warned of the difficulty in interpreting the data. However, the number of people reporting the same reactions without knowing they were one of many, were too powerful to ignore. I was astounded by the results of this study that took place years earlier.

Another document that Mary showed me was equally disturbing. It was an article written by H. J. Roberts, MD and published in the Journal of Applied Nutrition. Dr. Roberts collected data from surveys of respondents and used information in his article from 551 persons he labelled "aspartame reactors" because of their quick responses to aspartame (when it was removed from their diets and added back to their diets). Unlike the CFSAN study, Roberts' was able to interact directly and frequently with the participants in his study which allowed him to monitor their consumption of aspartame. He was able to establish what he believed was a causative relationship between adverse effects and aspartame consumption based on two things: "(1) the relief of complaints shortly after avoiding such products [that contain aspartame], and (2) their recurrence within hours or days of re-exposure [to aspartame], frequently inadvertent." I had only gotten through the introduction and I was already shocked at what I read. The next thing I read was even more appalling. Roberts quickly reviewed the three components that make up aspartame: phenylalanine (fifty percent), aspartic acid (forty percent) and methyl alcohol (ten percent). Methyl alcohol, which is more commonly known as wood alcohol, is toxic. Although Roberts did not discuss the components at any length, I made a mental note to research these three items and how they react with the human body. I continued reading, curious to see what Dr. Roberts would reveal about aspartame's effect on the people in his study.

The participants in Roberts' study consisted of 406 females (seventy-four per cent) and 145 males (twenty-six percent). The average age was forty-three, but they ranged in age from two years old to ninety-two years old. More than 250 people in the study said they used aspartame as a sugar substitute because of a weight problem. The majority of the participants (almost 350) said they stopped using aspartame completely when they first thought it was responsible for their symptoms. Of those people, ninety-six percent reported that their symptoms improved and seventy-four percent claimed that their symptoms disappeared entirely. These results made me believe that the vast majority of people who believe their symptoms are caused by consumption of aspartame are absolutely correct.

Even if a few people who linked their problems to aspartame were wrong, the overwhelming number of people who, to me, were obviously being effected by the artificial sweetener should not and could not be ignored. It seemed to me, after reading Dr. Roberts' article, which went on to give descriptions of several case studies, that aspartame was affecting people in ways we couldn't imagine. And, even if some people could use the product without apparent problems, the multitude of people sensitive to it was reason enough to investigate the symptoms further. Yet, the FDA appeared to believe aspartame was safe, placing no credence in the complaints and symptoms documented independently and suffered by thousands of victims. The FDA gave the public no warning, and the chemical was allowed into our food supply. Why hadn't I heard about this before now? Why was I only now reading these reports and studies on the dangers of aspartame? The impact this sweetener had on these people whom I read about and on my own thyroid gland was appalling. Although my suspicions about my own illness had now been confirmed, I vowed to keep on researching the topic over the next days and weeks.

•••

Time was quickly passing. Three months had already passed. I felt better and had no signs of Graves' disease. I was ready to see my physician and to confess what I had been doing and not doing. I really wanted to tell him what I'd found out about aspartame. I wanted him to know that I was on a natural food and vitamin-supplement program. But I waited, gathering more and more information, making sure I was confident of all of my facts before I would tell him that I hadn't been taking my thyroid medication and that I had found the cause of my thyroid symptoms. "The proof is in the pudding" was the adage I adopted. My good health had to tell him something. My thyroid was normal. My blood pressure, cholesterol, and heart rate were normal, too. I was no longer experiencing those intense migraine headaches and my mood swings were gone. My blurred vision and night blindness had slowly subsided, yet the retinal tearing in both eyes remained. My equilibrium was normal again, and my menstrual periods were right on schedule with no spotting in between.

Finally, the moment I'd been waiting and hoping for had arrived.

Sweet Poison

Carrying a huge briefcase of papers, I arrived for my regular doctor's appointment and stretched out my arm for one last conclusive blood test. I waited for the results.

"What's the verdict, Nurse?" I asked one last time.

"Normal," she replied.

Never had I felt such confidence. "I would like to see the doctor in his office, if I could, please."

Making some notations on my file, she excused herself and went to find him. Returning moments later she said, "Come this way."

She escorted me into the doctor's office. Despite my new found confidence my palms were sweaty and I was nervous as I trudged along hugging my briefcase. I had taken a chance going against the doctor's medical expertise. He didn't even know for the past three months I hadn't been on any medication. "He can really tear me down for this," I murmured, thinking out loud. "Do I want to set myself up for disapproval? I must."

"Did you say something," the nurse asked, turning around to look at me.

"No, nothing at all."

"That way. Last door on your right," she said pointing down a short corridor. She turned and walked away, leaving me to face the doctor on my own.

I found the doctor sitting at his desk, looking down at a file as I timidly entered his office. I guessed it was mine. I took a very deep breath. I walked toward him. I nervously stumbled over my own feet, but disguised the blunder by falling into one of his oversized leather chairs. We sat in silence and stared at one another for what seemed an hour, but was only a few minutes on the clock I watched on the wall. He idly thumbed through my medical file. We each waited for the other one to open the discussion.

He broke the silence. "Your thyroid has returned to normal and has done so in record time," he said. "How very unusual it is to encounter someone who has recovered from a case of Graves' disease," he added, looking at me uneasily. "But you are indeed an unusual patient." If only he knew how unusual! As he continued to speak so highly of my successful recovery, I began to think of the right words to tell him what I had kept secret over the past three months. I wanted to share what I had done, but I was uneasy confessing my insubordination.

Instead of speaking, I lifted dozens of vitamin bottles from my oversized purse and neatly arranged them on one side of his desk, each one rattling and clanking with tablets and capsules. On the other side of the desk I stacked copies of articles Mary Stoddard shared with me and piles of the documents on aspartame I uncovered through my own research. Finally, I spoke. "Each of these documents verifies the dangerous effects from the aspartame in NutraSweet. Dangerous effects on people' I stated. "Effects ranging from migraines to PMS and menstrual problems - everything I experienced."

As he reviewed them, his response was uncharacteristic for a doctor, but I loved it: "Boy, this aspartame is bad shit!" he blurted out.

Ha! I smiled.

As he continued to scan the papers, he acknowledged the information with concern. "You know," he admitted, "I learned very little nutrition in medical school and I am completely unaware of any problems with NutraSweet's aspartame products. I'm really much too busy to keep up with the countless changes developing in pharmaceuticals," he continued.

"So, I subscribe to an information service that supplies me with the latest information in the industry, by the industry. I've never received any information about aspartame. Why haven't I heard about this before?" he asked.

I smiled again and replied, "That's the question all of us should be asking. I have my theories," I went on. "And I think pockets run too deep for simple answers. However, I am convinced it's the sugar-free stuff that made me very sick."

My confidence building with his positive response, I handed him an article written by Lendon Smith, MD "Dr. Smith is a noted paediatrician and author. He has been featured on numerous television and radio interviews concerning contemporary health issues."

"I've seen him on TV," the doctor said.

I nodded. "Dr. Smith has stated that eighty-five percent of all complaints registered with the Food and Drug Administration concern reactions to aspartame, the sole ingredient in NutraSweet, and there have been four known deaths in FDA files listed under aspartame symptoms.

"It's funny," I went on. "Symptoms of using products with aspartame run the gambit. Over ninety-two different symptoms are documented- ninety-two! Some people get headaches, some nausea and vertigo. Others suffer with insomnia, numbness, or blurred vision. One person may become hyperactive and another suffer extreme fatigue. Dr. Smith listed neurological, dermatological, cardiac and respiratory problems associated with the use of aspartame. He reported symptoms of food sensitivities, low blood sugar, Alzheimer's disease, chronic fatigue syndrome, amalgam filling disease and methanol poisoning. Everyone agrees that their symptoms vanish when all aspartame use is terminated."

I shifted about in my chair, making the doctor shift around in his.

"Even though aspartame is about one-hundred-eighty times sweeter than sugar," I continued feeling like I was the doctor, and he the patient, "it isn't worth it when you find out what's in it. There's the amino acid phenylalanine, which is a known toxin to the brain and causes seizures. There's the amino acid aspartic acid, which can cause brain damage, especially to a developing brain. And then there's methanol, which creates eye problems and, when heated, turns into formaldehyde-embalming fluid, an obvious toxin-unless you're dead!"

I pushed on, hoping to get some sort of reaction out of the doctor. "Aspartame does not affect everyone equally. Common to most chemicals in solution, aspartame dissolves in the body. Its toxic by-products can then be deposited anywhere. Usually in the weakest spots, like my thyroid gland, making it even weaker and eventually breaking the system down.

"This is one reason why there are so many different symptoms related to using products with aspartame," I continued lecturing. "The chemical by-products dissolve in the blood and travel throughout the entire body. Who knows where the wastes will be deposited? They can go to the brain and cause headaches. If they go to the eyes, the results can be blurred vision, blindness, and in my case, retinal tearing. Toxins deposited in the pancreas can cause hypoglycaemia and diabetes. And if they go to the thyroid, of course, you can get Graves' disease.

"Dr. Smith asked in one of his articles, 'Why is the scientific community having trouble evaluating these reports? Might there be a substantial loss of revenue involved?'" I shoved Dr. Smith's article across the slick desktop.

"You know, Doctor," I stated, "if I unknowingly eat or drink anything with aspartame, I have an instant reaction. This has happened twice in the past three months. I took the kids to the cafeteria a couple of weeks ago. We all grabbed a cherry gelatine. I didn't realise it was diet – it wasn't labelled 'sugar-free'. But, boom! I had an immediate reaction. I got one of my bad migraines within minutes. I had to sit in the cafeteria holding my head until the pain subsided. Another time, my girlfriend served me lemonade sweetened with aspartame. Both times, my heart immediately started beating faster, I developed headaches, started to sweat, and got very, very hot. I had trouble breathing and suffered an asthma attack. Exercise induced asthma, I believe you call it. But I wasn't exercising. I was sitting still. Within minutes, I knew I'd fallen prey to hidden aspartame," I concluded. "I'm like a barometer for aspartame-laced foods," I said, jokingly. Maybe some humour would lighten the mood, I thought. The doctor didn't crack a smile.

Okay, then. Time for a deep breath. I stared at him. Was he going to say anything? He seemed genuinely interested at first and now, nothing. I'd better keep going before he started attacking me or calling me crazy or something. I handed him another stack of documents Mary had given me.

"Doctor, what about this?" I pointed to one of the articles. "Aspartame was first discovered in the late 1960s as an ulcer drug. By accident, a chemist discovered that the drug was sweet. Voila, aspartame was born! Corporate bigwigs eventually created the NutraSweet Company, and the campaign to revolutionise the sugar-free diet industry was on its way to the top.

"Scientists for and against this new sweetener immediately began an extensive research campaign. In nineteen sixty-nine, Dr. Harry Waisman fed his laboratory monkeys aspartame. One died after three hundred days on the stuff, while five others had grand mal seizures. According to Dr. Waisman, the results of this study were deleted when aspartame testing was first submitted to the FDA.

"The following year, in nineteen seventy, the FDA banned the use of cyclamates, and the safety of saccharin was seriously questioned. It was perfect timing for the introduction of a new sweetener. Aspartame seemed to fit the bill. By nineteen eighty-four, seven million pounds of NutraSweet had been consumed by over one hundred thousand people in the United States. The next year, consumption rose to sixteen million pounds. That same year, consumers started reporting side effects. By nineteen eighty-seven, four deaths associated with NutraSweet had been reported to the FDA.

"Whether or not I or anyone else chooses to use aspartame is a matter of personal preference," I said as my energy built. "But you can't make an educated decision if the negative side of this sweetener is kept hidden from the public. It's only fair when you choose to buy something that you know its good and bad points. Especially if it affects your health. I was using NutraSweet products with aspartame when I got sick. When I stopped using the stuff, I got well. It's as simple as that! I have to connect aspartame with my Graves' disease."

"Well..." he sputtered.

"No, there's no well. I definitely believe aspartame is responsible for my blinding headaches, along with provoking my recent eye problems. I haven't had one headache since I stopped drinking all diet drinks with aspartame.

"And researchers are aware that aspartame disturbs thyroid function," I added. "The pituitary gland uses serotonin as its signal to tell the thyroid what to do. Aspartame inhibits the formation of that neurotransmitter. The signal then does anything it wants." I felt this doctor really needed to know this! After all, he was the one who had wanted to permanently destroy my thyroid when there was no need.

"Serotonin is the neurotransmitter that determines sleep patterns, too. The aspartame inhibits that sleep signal. No wonder I had trouble sleeping."

Suddenly, something hit me. I looked at him. "Why don't you know any of this, Doctor?" I questioned in frustration. "As a doctor, you should know the background on the different chemical medications you prescribe for your patients. Medications can have dangerous reactions with other medicines and with the food chemicals people eat and dunk every day, all day long. Have you ever thought about this?"

He shifted uncomfortably in his chair, but I was in no mood for excuses. I rushed on.

"You, of all people, Doctor, know any drug, prescribed or not, has some sort of negative side effect associated with it. Just scrutinise the medication you have me on"- or had me on, but he didn't know this yet-"You told me in the hospital that the medication could destroy my immune system, remember? If I were still using aspartame, how do you think my medication would react with it? My guess is not very well. I think that I'd be even sicker."

Then I asked him the million-dollar question, "Doctor, would your diagnosis of Graves' disease have been different if you knew all this when you first met me in the hospital?"

He stiffened in his chair as he seemed to reflect on my question. Stealing a quick breath as if to speak, he shifted about and finally said, "This is very interesting, Jan." At least he knew my name by now.

Well, he heard me out, but, for me, this was hardly enough. I moved to the primary reason I had asked for this meeting. At this point, I felt comfortable offering a full confession. Well here goes!

"I think you should know that I have not taken my thyroid medication over the last couple of months." That did it! His cheeks became flushed, changing from pale pink to flaming red as his entire face swelled like an inflated balloon. The silence in the room abruptly broke as he shoved his large leather chair aside with an unexpected bang. For a moment, I believed he was going to jump over his desk, place his expensive hands around my less-expensive neck, and terminate all my worries about ever recovering from any disease again!

But as swiftly as his mood inflamed, he gathered his composure and relaxed his stance. Taking an incredibly long deep breath, he slowly spoke in a compassionate tone. "The information you have brought me is very interesting, but" – he strung the word out until I thought he was waiting for the music to start – "but," he repeated, "you were foolish to act upon it without consulting me first." In a parental tone, he reminded me that he was the doctor, not me. He said some other stuff, too, but by now I was not paying as much attention to what he was saying as how he was saying it.

As the energy in the room continued to swell, he shouted, "You don't know enough about what you're trying to do here. You shouldn't be taking this many vitamins. You don't know enough about Graves' disease, medicine, or aspar what-ever to be making these types of decisions. I don't even know that much about any of this. I do know that high dosages of these... things can do you more harm than good." He reached over and grabbed one of my vitamin bottles, looking at it as he held it high, shaking it with the passion of a politician making a campaign speech. Rattling the bottle, he said, "Leave the medical decisions to me!"

I sat frozen in my chair unable to respond. Glancing at his diplomas on the wall, I felt a fool for a split second to mistrust the omnipotent medical kingdom. Then I remembered how my headaches ceased and the dumps of my hair stopped falling out when I halted all medication and aspartame. "I can't forget the unsolved mystery of Graves' disease - no known cause but quick surgical solutions damaging my body." My throat tightened at the thought of swallowing that radioactive cocktail. The last few months reeled through my mind as I sat motionless in his overstuffed chair. I rehashed all the emotions I battled throughout the duration of this illness: the stress on my marriage, my constant yelling at the kids, my inability to function because of major migraines. I remembered all the useless conversations with doctors, the hours I spent with Steve Fugua learning how to regain nutritional common sense, how I learned to avoid hidden pitfalls like aspartame consumption, and how to protect my children from mislabelled toxic foods. I recalled the pages and pages of literature I read about aspartame and the testing that resulted in holes in the brains of laboratory mice, and the dead foetuses. I remembered also the congressional reports from medical doctors and researchers warning that aspartame might be dangerous to public health as well as aspartame's original discovery as an ulcer drug, not a sweetener. Spinning through my mind were scenes of pulling off the road as I drove home from campus suffering through another episode in which pinhole-sized points of light obstructed my vision. I could feel the nausea when the doctor first told me I had Graves' disease. The days and weeks I spent educating myself on thyroid function and how aspartame impacted the endocrine system marched past in my memory. I recalled how angry I was when I realised the truth about aspartame. No one can teach the courage it took to stand up to death. My memories furnished me with the strength to smile as I stood up to leave. "Well, I want to thank you for hearing me out."

Staring at me, the doctor backed down a little as he said, "Will you please let me know what you are doing next time before you do it?"

I met his eyes but didn't answer the question. Instead, I scooped all my vitamins into my purse with one quick swoop of my arm. "Thanks, Doctor," I said and walked out, my back stiffened.

He cried after me, "But it will come back! No one can cure Graves' disease. Listen to me, Jan. I care about you. It will come back!"

TIME OUT

I shifted down to second gear as I pulled up to a traffic light. Well, I no longer hare a doctor, I said to myself as I idled at the stop light, the vibration of my red Trooper rattling the bottles of vitamins I'd thrown into my bag. I really wanted to find someone in the medical community who would support me. What if I do have a relapse? "A relapse?" I cried out as the light turned green. I turned right and headed down a narrow road towards home, my thoughts bouncing back and forth. "Why do I think I'll have a relapse? Just because the doctor says so?" But I couldn't help myself; I was one of those people raised with the impression that when a doctor says something, because he's a doctor, it's inevitable. I paused and reflected. "Not this time. Not me," I stated with conviction. "I refuse to give in to this disease."

I knew aspartame caused my Graves' disease. But the doctor's foreboding predictions planted seeds of insecurity within me. What if the Graves' does come back? What mill I do? Who can I turn to for medical help? You know, I'll show the doctor I know what I'm doing. I'll prove aspartame made me sick every day.

Pulling into the driveway dodging basketballs and big wheels, I vowed to expose the dangers behind aspartame. Sitting in my car under a canopy of trees, I pledged to make daily trips to the library to research everything I could find on artificial sweeteners. I wanted to help Mary Stoddard continue researching case histories and promoting anti-aspartame campaigns. Any day, I expected the truth about aspartame to hit the news. "How can people be kept in the dark about this for so many years?" I wondered aloud. "The shoddy science. Holes in the brains of laboratory mice. Thousands of complaints registered with

55

the FDA. People have to be made aware of these things."

Over the next few weeks, I gathered as much research information on aspartame as I could. One afternoon, while sitting in the library, I mulled the facts over in my head. The research I had done thus far revealed that over 200 million Americans and another 200 million people world-wide regularly consumed products containing aspartame, supporting a multi-billion dollar business. How could they not? I thought. NutraSweet was approved in over one hundred countries and I came across a list of over 5,000 products that contained it.

I shuddered to think that the list would keep growing

Just because NutraSweet has such widespread sales doesn't mean its a safe product, I thought as I sat at the large wooden table hidden in the depths of the library. The information I'm uncovering proves that aspartame s safety has been questioned since it was first discovered.

I came across information on the origin of the artificial sweetener and, even though I knew some of it, I wanted to know the whole story, so I read carefully. In 1965, James Schlatter, a chemist for G.D. Searle Company, was testing an anti-ulcer drug in his corporate laboratory. While he was experimenting with protein combinations, the beaker bubbled over onto his hand. Schlatter happened to lick his fingers to pick up a piece of paper that had fallen to the floor. "Ah, sweet sensation," he is reported to have commented, "this anti-ulcer drug is sweet." So, from this anti-ulcer medication an artificial sweetener was born. The thought of an anti-ulcer medicine being used to sweeten over 5,000 products still sent a chill down my spine.

I continued reading. This chemical white powder revolutionised the contemporary food industry. Its creators claimed, "It takes away the bittersweet taste of diet drinks." The facts again appalled me. The only thing that I feared more than what I was reading about, was that most Americans have never heard of the numerous studies that question the safety of aspartame, the registered complaints, the brain tumours in mice, and all of the other horrifying things I was learning. Although people like Mary Nash Stoddard have worked doggedly to spread the message and make consumers aware of the dangers of aspartame, I felt a need to join that effort. I hoped to have an opportunity to speak out and educate others about this sweet poison.

Since 1987, Mary Stoddard has been consulting daily with scientists, educators and professors, researchers, victims and their families, physicians, therapists, pilots, and people from every segment of society who have had negative reactions to aspartame. Since meeting Mary, I did what I could to help her in these consultations and research efforts. I felt fortunate that we were getting to know one another and sharing ideas.

"You know, Jan," Mary said one day over tea, "I defeated a near-fatal blood disorder in the mid nineteen eighties that I sourced to aspartame. I vowed then to do something about this sweet poison."

That's just what she did, too. She solely operates the national Aspartame Consumer Safety Network (ACSN) and international Pilot's Hotline.

Mary and I, both victims of aspartame poisoning, became two women on a mission. Neither she nor I would stop exposing the truth about the dangers of NutraSweet and aspartame until it became common knowledge.

One morning, Mary telephoned to share an unexpected idea with me. "I've decided to sponsor a conference at the Dallas/Fort Worth International Airport," she said with enthusiasm. "I'm going to present facts and scientific research on the dangers of aspartame. Will you speak about your miraculous recovery from Graves' disease?" she asked.

"Of course, I will," I quickly replied with enthusiasm, my opportunity to speak out was coming sooner than I expected. "Anything to help get the word out. Talking about what I've been through helps me get stronger. I'd love to be there."

Mary personally funded the conference and paid for all advertisements in the local newspapers and on the radio. She even rented a conference room at the Harvey Suites Hotel. Admission was free. That night I was nervous, but prepared and anxious to tell my story.

Mary began. She was an eloquent speaker and spoke about the history of aspartame. It captured the attention of the 150 people in the audience. After Mary spoke, I shared my story. Beginning with my symptoms, I led up to my Graves' diagnosis and the doctor insisting I have my thyroid removed. Then I discussed how I'd resisted and instituted my own program of recovery through natural foods and vitamins. Then, I explained what aspartame really is.

"Aspartame is the scientific name for the brand names NutraSweet, Equal, Spoonful, and Equal-Measure," I said in a professorial tone. "NutraSweet is considered a food additive. Equal, a sugar substitute, contains NutraSweet, dextrose with dried corn syrup, silicon dioxide, cellulose, tri-basic calcium phosphate and cellulose derivatives.

"A white crystalline powder, aspartame is one-hundred-eighty to two-hundred times sweeter than sugar," I continued, gaining momentum. "But, what is aspartame actually? What makes this product so questionable? Aspartame is made up of fifty percent phenylalanine, an essential amino acid found naturally in foods such as milk and bananas, forty percent aspartic acid, also an essential amino acid found naturally in foods, and ten percent methyl alcohol (methanol), a toxic, colourless, flammable liquid, and the precursor to formaldehyde. Methanol is toxic to humans. Bottom line, methanol is not food and has no business being consumed in any foodstuff. Period."

I saw many eyes widen and some people shake their heads in disgust. I pushed on, determined to get the whole ugly truth out. "Even though phenylalanine and aspartic acid occur naturally in foods, aspartame is not natural. The phenylalanine and aspartic acid in aspartame are man-made, laboratory replicas of the real thing. The manufacturers of aspartame would like you to believe aspartame is all natural, but it's not natural in any sense!

"Think about this," I said, my voice rising. "It would cost a fortune for aspartame manufacturers to extract these two amino acids from natural foods. That's a lot of milk and bananas) The actual sources of phenylalanine and aspartic acid in aspartame are well-kept secrets."

I felt proud to be standing there, revealing deceptions that affected so many people's lives. "Richard Wurtman, MD, director of the Massachusetts Institute of Technology (MIT), stated in nineteen eighty-seven, 'NutraSweet is a synthetic compound. We should not try to imply that synthetic compounds have the same fate in the body as naturally occurring foods. The advertising is deceptive. The phenylalanine in NutraSweet zips right into the brain because you don't have the other amino acids to block its entry.' Ladies and gentlemen, Doctor Wurtman is right."

I paused for a moment, took a deep breath, and spoke in a slower, more deliberate tone. I didn't want anyone to miss what I had to say next.

"Since nineteen eighty-one, the Deerfield, Illinois-based NutraSweet Company has been the only company patented to sell aspartame in the United States," I told them. "It is a wholly owned subsidiary of Monsanto Chemical Company. Ajinomoto Company, Inc. is the Japanese licensee of the original aspartame patent. However, in

December nineteen ninety-two – next year – NutraSweet's patent will expire, opening the way for world-wide competition. At that time, anyone will be able to manufacture aspartame."
Even reciting facts that I had known for some time still shocked and bothered me.
Angry now, I added, "Even though aspartame is manufactured in Japan, the Japanese don't use it. They use stevia, a plant from South America, in their diet cola. They have for over twenty years."
Mounting a chart for the audience to see, I explained, "Some foods have aspartame plus other artificial sweeteners in them, such as saccharin. Now this is a dangerous proposition! Primarily because of the aspartame content. When NutraSweet first came on the market, Doctor Jim Bowen, MD, a private family practitioner, researched the combination of aspartame and saccharin and determined that, together, both aspartame and saccharin chemically combine to reach temperatures above boiling point. So, products with both artificial sweeteners are capable of scalding your bladder!" Some audience members gasped. "Doctor Bowen just sat down one day and charted his findings on how toxic aspartame really is," I said, angrily pointing to the chart which contained Dr. Bowen's own notations and handwriting. The disgusted reactions of the audience didn't surprise me at all, since they mirrored my own reactions when I first saw the chart. It represented the most graphic portrayal I had ever seen pertaining to the damaging intensity aspartame has on the body.
I continued, "Dr. Bowen wrote, 'Examination by numerous physicians will usually indicate that there is nothing wrong with the patient. They will be diagnosed as hypochondriac. For those affected, their world crumbles about them mysteriously. Meanwhile, neither they nor their physicians have an inkling of what's going on.' This definitely struck a chord with me considering my illness."
Mary handed out copies of a United States Food and Drug Administration (ERDA) Department of Health and Human Services report while I continued to explain that aspartame accounts for over 85% of the adverse reactions to food additives reported to the FDA. "Many of these reactions are very serious," I stated. "They include seizures and death as disclosed in this February nineteen ninety-one report. Here are just some of the symptoms related to aspartame that are listed in the report: headaches/migraines, seizures, numbness, dizziness, nausea, muscle spasms, weight gain, depression, irritability, insomnia, hearing loss, breathing difficulties, blurred speech, tinnitus,

memory loss, rashes, fatigue, tachycardia, vision problems, heart palpitations, anxiety attacks, loss of taste, vertigo and joint pain." The audience was visibly stunned by what they were hearing.

"According to researchers and physicians studying the negative effects of aspartame," I continued, "the following list cites chronic illnesses known to be affected, if not triggered, by aspartame: brain tumours, epilepsy, Parkinson's disease, mental retardation, birth defects, diabetes, multiple sclerosis (MS), chronic fatigue syndrome, Alzheimer's disease, lymphoma, fibromyalgia, and last but not least – Graves' disease!

"So does aspartame cause Graves' disease?" I asked rhetorically. There were loud reverberations in the audience. "Why didn't my doctor know this? He was so positive about destroying my thyroid gland, he never considered researching the cause of my disease, a much more attractive alternative to permanent destruction.

"How could my doctor recommend something so drastic as permanently destroying my thyroid gland when he didn't even know all the facts?" I questioned. "He not once mentioned nutrition as a part of my recovery." I paused and asked, "Has something like this ever happened to you?"

I paused another moment, allowing the audience to reflect on my question. Again murmurs rose.

"Instead of accepting what my doctor recommended," I continued, "I'm glad I took the time to question everything. Something was left out of the equation, and I had enough common sense and raw guts to find it. And I found it, all right! The cause of my Graves' disease was aspartame poisoning," I announced.

"My Graves' disease would be more accurately labelled aspartame disease," I proclaimed as I finished up my speech. "I suspect many more people with diseases and or symptoms similar to mine are merely reacting to chemicals concealed within their foods. Chemicals like aspartame."

Much to my surprise, at the close of the conference, Mary walked to the podium and boldly asked, "Is anyone in the audience from the NutraSweet Company? Would you stand up, please?"

What is she doing? I asked myself in horror.

Mary had told me earlier, an all-knowing grin stretched across her face, that she didn't know how they found out about her speaking engagements, but the NutraSweet Company always had a representative present wherever she spoke. Since the NutraSweet

Company still held an exclusive patent on aspartame, any negative comments aimed at their product were taken personally. However, I didn't expect Mary to point out the NutraSweet executive and give him or her the floor. An attractive woman dressed in a fetching tailored suit meekly raised her hand and rose.

"I am Sue Ross," she stated with poise. "I am the manager of public relations for the NutraSweet Company."

Like the rusty throttle of an old jeep, I choked up. I tried to regain control of myself as I forced a deep breath and gasped for air. The NutraSweet Company sent their manager of public relations to our lecture?

"I have never been to Texas," she explained. "But my stay is brief, as the NutraSweet Company has flown me to Dallas on the corporate jet exclusively to attend this conference. Shortly after the forum, I will fly back to Illinois."

After a brief moment of silence, she clarified, "NutraSweet's corporate office is located in Deerfield, Illinois."

She was a striking and poised woman who handled a difficult situation adroitly. Despite myself, I was impressed with her.

She began speaking, contradicting everything Mary and I just proclaimed about aspartame. She referenced statistics from their corporate research team of MD's and Ph.D.'s. I quickly became irritated with her rebuttal, which seemed rehearsed to me. However, because this wasn't the place for a debate, I struggled to keep my mouth shut. I fixed my gaze on Mary.

"Mary," I whispered, "how did they know we were speaking? You didn't advertise in Illinois. Why do they even care about us if their product is so flawlessly safe?"

"I don't know, Jan," Mary answered. "But I told you this always happens when I speak. Don't worry though, they always disappear by the end of these conferences. You'll see."

Indeed, Ms. Ross mysteriously vanished at the close of our speeches.

•••

A few weeks later, I sat at the kitchen table, huddled in my husband's terry cloth robe. Since my hospitalisation, I had procrastinated dealing with the disenchantment of my marriage. Staying healthy and preventing a relapse was my most important objective these days. Maybe time will heal old wounds, I convinced myself I'll worry about it later. For now, I want to concentrate on doing something about aspartame.

It was nearly seven o'clock on a Saturday morning, and the boys and Chuck were still asleep. I got up and poured myself a cup of coffee. Warming my hands around the steaming cup, I pulled a stack of papers out and laid them in front of me on the kitchen table. They were articles I had collected, but not yet read, from a recent visit to the library and another meeting with Mary. It seemed a perfect, quiet early morning hour to catch up on my aspartame research.

The first article I picked up was by James Erlichman, consumer affairs correspondent for the British newspaper, The Guardian. The article stated that the British government received a dossier of evidence in the late 1980s on the dangers of aspartame. The dossier alleged that the potential dangers of the artificial sweetener were concealed. The dossier also alleged that tumours were removed from laboratory animals fed aspartame and some animals that had actually died were miraculously "restored to life" for the laboratory records.

I felt a chill as I read this. Pulling my knees up, I gulped some coffee, then wrapped my arms around my legs for warmth before I continued reading.

The dossier was compiled by Erik Millstone, a lecturer at the Science Policy Research Unit at England's Sussex University and author of two books on food additives. Millstone based his study on thousands of pages of evidence, most of which he obtained through the United States Freedom of Information Act.

Before NutraSweet was approved in the UK in 1983, a warning label was agreed upon by the Committee on Toxicity, a British government panel. A Ministry of Agriculture spokesman said that adopting the warning was voluntary. The committee held no power to force manufacturers to display warnings.

That's great, I thought sarcastically. Warning labels that no one has to put on their products. I read on.

The British newspaper reported that aspartame (NutraSweet) was first approved in the United States for unrestricted use in July 1983 by Arthur Hull Hayes, FDA Commissioner. Despite objections from his own scientific board of inquiry and numerous independent researchers concerned that aspartame caused brain tumours, Dr. Hayes overruled their warnings.

Shortly after he approved NutraSweet, Hayes resigned from the FDA to join the developer and manufacturer of aspartame's public relations firm, Burston Marsteller. According to the article, however, an inquiry into Dr. Hayes' actions found no impropriety. Dr. Hayes, according to Erlichman's sources, had a reputation of being a man who believed FDA approval for new drugs and food additives was too slow, due to the amount of information the FDA required.

The confusion, misleading information and half truths that seemed to surround aspartame (NutraSweet) astounded me. I sipped my coffee before returning to the paper.

Also, according to the article, the Ministry of Agriculture and the US Department of Health never revealed what evidence NutraSweet's approval was based on. In the UK, those were matters of "commercial confidence," not public knowledge. I was stunned to read that neither government did testing of its own, but relied solely on data submitted by the company's scientists.

At that time, the US Department of Health announced that, based on the evidence available, NutraSweet did not pose a health hazard, but admitted people do suffer "idiosyncratic reactions" to food additives. The NutraSweet Company said it was satisfied with the safety of its product, although "a small group of consumers" could theoretically be sensitive to it. Like me, May, and the thousands of people who have registered complaints? I wondered.

"It's not right," I whispered to myself as I got another cup of coffee. "This aspartame thing just isn't right." I bundled the papers into a neat stack, and with hot cup in hand, made my way to the den sofa. Placing my coffee on the table, I sank into the pillows to continue my reading. Thumbing through the pages, I picked up a 1974 report written by the late Dr. Adrian Gross, Ph.D., former FDA toxicologist. Dr. Gross was employed by G.D. Searle to investigate the safety of aspartame.

Dr. Gross was present at the April 8-9 and July 10, 1976 joint hearings held by two subcommittees of the United States Senate. Senator Ted Kennedy, FDA Commissioner Alexander Schmidt, MD, and a task force which Commissioner Schmidt appointed were present.

Dr. Gross testified again on the Congressional Record in November 1987. In that Congressional Record, Dr. Gross stated, "Through our efforts, we have uncovered serious deficiencies in Searle's operations and practices which undermine the basis for reliance on Searle's integrity in conducting high quality animal research to accurately determine or characterise the toxic potential of its products."

He went on, "Searle has not met the... criteria on a number of occasions and in a number of ways. We have noted that Searle has not submitted all the facts of experiments to FDA, retaining unto itself the prohibited option of filtering, interpreting, and not submitting information which we would consider material to the safety evaluation of the product.

"Some of our findings suggest an attitude of disregard for FDA's mission of protection of the public health," he continued to state, "by selectively reporting the results of studies in a manner which allays the concerns of questions of an FDA reviewer. Finally, we have found instances of irrelevant or unproductive animal research where experiments have been poorly conceived, carelessly executed, or inaccurately analysed or reported."

A public board of inquiry was established to investigate the safety of aspartame in 1979. A year later, the board voted two to one that aspartame was not safe for public consumption. The board was particularly concerned about one of NutraSweet's by-products, Diketopipefazine (DKP), a chemical by-product directly linked to brain tumours.

The article continued, saying that an in-house team of FDA scientists examined the aspartame data. Three out of six members advised against approval due to the developing brain tumour issue. Dr. Hayes overruled the recommendations against aspartame and approved the chemical as a "table top" sweetener in July 1981. Two years later, in July 1983, Dr. Hayes granted full approval for aspartame in diet drinks. Two months later, the British Ministry of Agriculture followed suit.

I sat there, looking out the den window in a daze trying to comprehend all the information I had just read. How could these people, supposedly there to protect the public, allow products with aspartame to be on the market?

I mulled it all over in my mind for quite a while. I was so caught up in my thoughts I didn't even hear my son Sean enter the room.

"Mommy, I'm real, real hungry," he announced.

My startled look made Sean jump. I looked at his sweet face and lamented that I had ever given him food or pharmaceutical products that contain aspartame. Hugging him to me, I promised myself I would put all I'd learned to work by feeding my innocent, impressionable children the best and freshest natural products.

"How does toast, scrambled eggs, and a big glass of milk sound?"

His huge smile was the only answer I needed.

US PRODUCTS SWEETENED WITH ASPARTAME

BREATH MINTS, CHEWING GUM
Bazooka – Sugarless; Blammo Sugar Free Bubble Gum; Breath Savers
Sugar Free Mints; Bubbalicious – Sugar Free; Certs – Sugar Free; Extra
Sugar Free; Hubba Bubba Sugar Free Bubble Gum; Rain-Blo Gum Balls
– Sugar Free

CEREALS
All Bran with Fibre; Fibre One; Sun Flakes

**MILK ADDITIVES, INSTANT COFFEES, TEAS & COCOA POWDERED
DRINKS**
4C Iced Tea Mix; Alba 77 Diet Shake Mix; Alba Sugar Free Hot Cocoa
Mix; Carnation Sugar Free Hot Cocoa Mix; Carnation Instant Breakfast
No Sugar; Country Time Sugar Free Drink Mixes; Crystal Light Drink
Mixes; General Foods Sugar Free International Coffees; Hill Bros Sugar
Free Instant Chocolate Coffee; Kool-Aid Sugar Free Mix; Lipton Iced
Tea Mix; Nestea Sugar Free Iced Tea Mix; Nestle Quik Sugar Free
Chocolate Milk Powder; Nestle Sugar Free Hot Cocoa Mix; Ovaltine
Sugar Free Hot Cocoa Mix; Ovaltine Sugar Free; Diet Sun Drink Mix;
Sunkist Light Sugar Free; Swiss Miss Sugar Free Hot Cocoa Mix; Swiss
Miss Sugar Free Chocolate Milk Maker; Sugar Free Tang; Sugar Free
Wyler's Drink Mix; Tetley Iced Tea Mix; Twinings Sugar Free Iced Tea
Mix; Various Store Brand Sugar Free Drink Mixes

**FROZEN and REFRIGERATED JUICES, MILK BEVERAGES, ICED
TEAS**
Borden Light Chocolate LowEat Milk; Citrus Hill Lite Orange Juice;
Land 'O Lakes Lowfat Chocolate Milk; Lipton Diet Iced Tea; Minute
Maid Light 'N Juicy; Nestle Quik Lite Chocolate Milk; Ocean Spray
Diet Juice Cocktail; Red Rose Sugar Free Iced Tea; Snapple Diet Iced
Teas; Snapple Diet Juices; Tetley Diet Iced Tea; Treesweet Lite Juice
Beverages; Welch's No Sugar Added Juice; Various Store Brand Sugar
Free Beverages

DESSERTS
3 Musketeers Frozen Sugar Free Bars; Bob's Sugar Free Hard Candies;
Breyer's Light Ice Cream; Carnation Creamy Lites Frozen Snack Bars;

Creamsicle – Sugar Free; Crystal Light Frozen Bars; Dannon Light Frozen Yogurt; Dole Fresh Lites; Edy's Frozen Dietary Desserts; Eskimo Pie Sandwiches, Bars and Desserts; Freezer Pleezer Sugar Free Juice Coolers; Fro Yo (Frozen Yogurt); Fudgsicle – Sugar Free; Golden Batch Sugar Free Cookies; Gourmet Light; Healthy Choice Sugar Free Gelatin; Jell-o Sugar Free Instant Gelatin Dessert; Klondike Lite; Life Savers Sugar Free Flavor Pops; Nestle Crunch Lite; Nutri/System Gelatin and Puddings Mixes; PET Light Sugar Free Frozen Bars; Popsicle Sugar Free Ice Pops; Royal Sugar Free Pudding and Gelatin Mixes; TCBY Sugar Free Frozen Yogurt; Weight Watchers Chocolate Mousse; Weight Watchers Frozen Fruit Juice Bars; Welch's No Sugar Added Fruit Juice Bars

SOFT DRINKS
Diet 7-Up; Diet Cherry 7-Up; Diet 50/50; A&W Diet Cream Soda; A&W Diet Root Beer; Barq's Diet Root Beer; Barq's Diet French Vanilla Cream; Barrelhead Diet Root Beer; Diet Boost; Diet Bubble-Up; Canada Dry Sugar Free Ginger Ale; Canfield's Diet Flavoured Soft Drinks; Diet Cherry Coke; Diet Citrus 7; Diet Coke; Cotton Club Diet Soft Drinks; Diet Orange Crush; Diet Rite Cola; Diet Dr. Pepper; Fresca; Frostie Sugar Free Root Beer; Hawaiian Punch Diet Fruit Punch Soda; Hires Diet Root Beer; Hubba Bubba Diet Bubble Gum Soda; Jolt Diet Cola; Minute Maid Diet Sodas; Diet Mountain Dew; Diet Mug Root Beer; Diet Pepsi; President's Choice Diet Sodas; Diet Quench; Diet RC Cola; Diet Cherry RC Cola; Schweppes Diet Ginger Ale; Schweppes Diet Seltzers; Seagram's Sugar Free Ginger Ale; Shasta Diet Sodas; Diet Slice; Diet Sprite; Diet Squirt; Diet Sunkist; Super Market Diet Sodas; Tab; Yoo Hoo Diet Chocolate Soda; Various Store Brand Diet Soft Drinks

MEDICINES, VITAMINS, LAXATIVES
Acetaminophen (Children's Chewables; Alka-Seltzer Plus Cold, Cough and Sinus Medications; Anacin Chewables (Children's); Bugs Bunny Children's Chewable Multivitamin; Centrum Jr Vitamins; Dimetapp Cold & Energy Chewable Tablets; Flintstone's Children's Chewable Multivitamins; Health Balance Children's Chewable Multivitamins; Lifeline Natural Fibre Laxative; K-Mart Chewable Animal Shaped Vitamins; Metamucil (Sugar Free); Mylanta Sugar Free Natural Fibre Supplement; Naturlax Natural Fibre Laxative; Pedia Care Children's Cold and Allergy Tablets; Scooby-Doo Chewable Multivitamin; Sunkist

Multiple Vitamins; Tylenol Acetaminophen Children's Chewables;
Tylenol Cold and Flu Medication; Various Pharmacy Brand Children's
Chewable Multivitamins

YOGHURT
Borden Light Nonfat Yoghurt; Colombo Slim Yoghurt; Dannon Light;
Light 'n Lively Sugar Free Yoghurt; La Yogurt Light; Penn Maid Light;
Yoghurt Prairie Farms Light Yoghurt; Weight Watchers Ultimate;
Yoplait Light; Various Store Brands

MISCELLANEOUS
Various Brands of Sugar Free Fruit Spreads and Jams; Maple Grove
Reduced Calorie Maple Syrup; Various Brands of Protein Shakes;
Various Brands of Sugar Free Whipped Toppings

HOLD ONTO YOUR
LAB COATS

C onsumption of even moderate amounts of aspartame during pregnancy... may produce a dramatic increase in the number of children born with diminished brain function.

Diana Dow-Edwards, Ph.D. SUNY Health Science Centre, Brooklyn, NY

The smell of worn books, dim lighting, and high ceilings felt homey to me. Every inch of wall space was covered with rows of books. The silence was so thick you could hear your mind mark thoughts like an old printing press. I loved libraries.

Never had I spent so much time poring over the printed word, books, journals, and federal registers than during my Graves' recovery. I became obsessed with researching aspartame. And, the more information I encountered, the more I wanted to find out.

Ever since that morning I had sat in the den looking at my son and realised how important nutrition was for growing bodies, I'd wanted to learn about the effects aspartame has on children. Once the truth could be exposed, adults would have the opportunity to choose for themselves whether to use the chemical sweetener or not. But children depended on adults to make those choices for them. I had to learn about this so I could make adults – parents – aware of the dangers that faced both them and their children.

As I delved further and further into research, I learned that foetuses exposed to aspartame could suffer catastrophic fates before they were even born. One of the most dramatic and unsettling research reports I found through my investigations was written by research scientist Diana Dow-Edwards, Ph.D., at the SUNY Health Science Centre in Brooklyn, New York in 1989. Dow-Edwards' study focused on pregnancy and foetal formation.

Dr. Dow-Edwards was granted $128,389 to analyse the effects of aspartame on the development of guinea pigs. She was asked to investigate evidence of mental retardation and developmental disabilities as a result of using aspartame during normal pregnancy. She evaluated the general development in the offspring of guinea pigs fed aspartame while in utero.

Since brain development occurs in the womb in both guinea pigs and humans, the results of her investigations were applicable to human development, too.

Dr. Dow-Edwards administered aspartame to pregnant guinea pigs during their entire pregnancy via their drinking water. Since phenylalanine has been proven to cause most of the toxic effects from aspartame and because guinea pigs metabolise phenylalanine more rapidly than humans do, she fed additional phenylalanine to some of the groups. She divided the guinea pigs into five groups: 1) one control group fed plain water; 2) one control group given 34 mg/kg/day (milligrams per kilogram of body weight) aspartame in their water; 3) a control group fed 34 mg/kg/day aspartame with phenylalanine in the water; 4) a group given 1,000 mg/kg/day aspartame in their water; 5) and a control group fed 1,000 mg/kg/day aspartame with phenylalanine in the water.

Ten offspring (five males and five females) in each of the five groups were studied. Litter data was collected. One pup from each litter was sacrificed on the day of birth so its levels could be used as a benchmark for the study. A blood analysis, liver and kidney function tests, a soft tissue inspection, and brain removal for the study of brain formation was documented on the forfeited pups. The remaining pups were extensively evaluated for reflexes, reactivity, activity, and visual perception. Both the adult guinea pigs and their pups were tested for brain function.

I couldn't help but cringe at the thought of what those poor little pups went through, but I had to read on. Dow-Edwards' research evaluated the ability of moderate doses and toxic doses of aspartame to harm pregnant mothers. Dow-Edwards monitored the blood levels of phenylalanine so she could accurately apply the results to human pregnancy. Her results were particularly important to women with liver problems because of their inability to metabolise phenylalanine and to phenylketonurics who are individuals born with the inherited inability to metabolise phenylalanine.

As I read about the effects of aspartame and children, I kept running across the word phenylketonuria, or PKU. Phenylketonuria appeared to be a very serious thing, yet I'd never heard much about it before studying aspartame. If its so dangerous to humans, I asked myself, then why isn't it mentioned more?

With this in mind, I put aside Dow-Edwards' reports for a few minutes to read more about PKU. I learned that every aspartame (NutraSweet) product carries the warning label: Caution: Phenylketonurics, contains phenylalanine, because phenylalanine causes mental retardation in individuals with phenylketonuria or PKU, and phenylalanine makes up 50% of aspartame. There are approximately 20 million carriers of the PKU gene at risk who do not know it. I was surprised that what seemed like a widespread problem was relatively unknown to the public.

Phenylketonuria is a disease characterised by an inborn error of metabolism of the amino acid phenylalanine. In affected individuals, the disease is present from birth. It is not clinically recognisable at first. Progressive mental retardation occurs from the age of a few weeks. Irritability and vomiting are early symptoms, and dermatitis may appear at five or six months of age. Affected children have fairer complexions than their unaffected siblings, and can become so retarded they often require institutional care. Victims also frequently develop epilepsy.

Phenylketonuria is inherited on a recessive basis, meaning both parents can be unaffected, yet carry the trait. There is a one-in-four chance of producing an affected infant in every pregnancy. In 1980, one out of every sixteen thousand live births were phenylketonuric in the United States.

I wondered how carriers of the PKU trait would react to high doses of phenylalanine in foods they consumed, like diet products. I continued turning the worn pages of the medical book as quickly as I could read them.

Treatment for Phenylketonuria involves a strict preventive diet which contains little to no phenylalanine – that means no aspartame! What would you do if you didn't know you had PKU? What if your two-month-old baby had PKU and you didn't know? What if you were pregnant and had no clue if the foetus had PKU? I shuddered at those thoughts, then realised that since I'm adopted, I have no idea what my genetic history is.

"No one has ever mentioned PKU to me before," I said half under my breath. I sighed as I stared at a bug crawling up a shelf of leather-bound medical books. I've never read an article on PKU nor seen one television talk show warning people to slow down their use of aspartame due to PKU risk, I thought as I stared at the insect. The hardworking bug successfully reached its goal of crawling behind an anatomy book. Its disappearance caused me to snap back to reality and I once again returned to the volumes of information before me.

I continued reading the PKU report which explained that many common foods such as milk and bananas contain phenylalanine because phenylalanine is a naturally occurring amino acid A small amount of phenylalanine is required for normal growth, but no more than the small amount found naturally in food. Diet products with aspartame don't count as a source of the "naturally occurring amino acid" because their source of phenylalanine is not a direct food source but a laboratory replica of the natural phenylalanine molecule. Plus, with over 5,000 products containing aspartame, the average consumer has a better than average chance of receiving too much phenylalanine. According to this information, the conclusion was, your diet, especially if you're phenylketonuric, must be monitored very closely. As a closing point, the article said that phenylalanine's effects are cumulative: they build in the body over time.

I sat back in my chair to digest all that I had learned. Mulling it over, I concluded that if a person got phenylalanine from natural food sources plus consumed many synthetic sources of phenylalanine, like aspartame for example, they would be getting too much phenylalanine in their daily diet, and resulting problems would begin to accumulate. To me, that was a scary thought.

Feeling satisfied with the amount of information about PKU that I now had, I returned to Dow-Edwards' research to see what she came up with. I sort of dreaded finding out.

And I was right to feel a sense of foreboding. Dow Edwards had concluded from her studies that using even moderate quantities of aspartame during pregnancy could produce a dramatic increase in the number of children born with diminished brain function. This study and its findings, I thought, were enough to warn the public to avoid all aspartame products, especially during pregnancy!

Weary from researching and feeling dismal over the things I was learning, I decided to pack up and head home. I needed a break, but I knew I would be back. The things I had learned about the negative effects of aspartame only made me want to know the whole story.

Not long afterward on a typical hot and sunny Texas Monday while driving to school, I noticed a billboard towering over the highway advertising a children's hospital in Dallas. The billboard depicted the "remodelled" hospital as a storybook house, the script boasting, "We have proudly added seven new stories to our facility."

That billboard struck me as odd. Was this really something to be proud of? Seven new floors of sick children? I didn't think so! In fact, I thought it was a depressing idea.

It got me thinking though. I considered how many children are contracting "adult" diseases these days: cancer, diabetes, MS, depression, eating disorders, seizures and all kinds of degenerative diseases like Graves' disease. It seemed to me that something different was happening that hadn't been a problem for children in the past; something was triggering an increase in disease syndromes. Again, I came to the conclusion that the abundance of chemicals, like aspartame, in our food must fit into the equation.

I turned in to the university parking lot, being careful not to spill my travel mug filled with coffee and pulled into my spot. I sat there thinking about that billboard. Then, I thought once again about my illness. If drinking diet soda could cause my Graves' disease, it was disturbing to imagine how sensitive children's little bodies must be to aspartame. And, even more worrisome to me was the thought that so many children are exposed to aspartame on a daily basis. Dow-Edwards proved its effects on foetuses. That was enough to frighten me.

I got out of my car and headed towards my office, still pondering. I walked slowly, thoughtfully unaware that people around me were in a rush. It wasn't until a student bumped into me that I lurched back to my senses and realised class would soon start. I hurried to my office.

Pausing to pick up my lecture notes and dump my unfinished coffee in the faculty sink, I shut the office door and dashed to class in barely enough time to beat the dock.

My students, appearing as dishevelled and rushed as I, were all in the mood to talk about current issues as opposed to their assignment on international politics in the former Soviet Union. "That's okay," I said, giving in. "Sometimes, current events are the best curriculum. As long as you learn something today, let's talk about what's of interest to you guys."

Since many people on campus were talking about the professor who'd gotten sick from artificial sweeteners, they knew about my illness and that I believed it was caused by aspartame. So, it didn't surprise me that the discussion led to aspartame. The first question that came up was, "If aspartame isn't safe, then why did the FDA approve it?"

"I ask myself that same question, too," I responded.

I explained to them that the FDA had known about the dangers of NutraSweet since the early 1970s, and why they didn't do anything about it was beyond me. I explained, "The FDA is aware of the research results from numerous independent scientists proving the dangerous effects of aspartame, yet they discount this research by doing nothing." With an exasperated sigh, I went on, "They seem to dismiss all the negative research results – research showing holes in the brains of laboratory mice, mammary gland tumours, brain seizures, dead and deformed foetuses, and so many more. Yet, the FDA acted upon the test results submitted by researchers on the corporate payroll. NutraSweet claims their research proves aspartame is 'safe for public consumption.' That's all the FDA seemed to need."

"Have there been any formal government reviews submitted proving aspartame unsafe?" asked one student.

"Definitely," I replied. "The FDA has a list that they put out this year called 'Symptoms for aspartame'." Always prepared, I had a copy of the list in my briefcase. I pulled it out and passed it around the classroom so all my students could look at it before I continued. I watched their horrified expressions as they read.

"Interestingly, the FDA flagged certain symptoms on the list, including two hundred and one symptoms reported for other neurological, one hundred seventy-four grand mal seizures, ninety-two other, eighty-four local swelling, thirty unspecified, twenty-six petit mal seizures, twenty-two nocturnal, ten simple partial seizures, four deaths, and four complex partial seizures. The distribution was based on five thousand four hundred twenty-two records and eight thousand and one occurrences of symptoms.

"Since the approval of aspartame, the FDA has received over five thousand complaints against aspartame, which comprises eighty-five percent of all complaints ever registered with the FDA. Prior to nineteen eighty-seven, four deaths associated with aspartame were recorded with the FDA. There are only six documented complaints filed against saccharin. A total of six complaints in over one hundred years of saccharin use."

"Four deaths have been associated with aspartame?" another student interjected, shocked. "I wonder how many more deaths that we don't even know about are linked to aspartame today?"

"That's a good point," I responded. "The number of deaths from seizures has increased in general. Paramedics will tell you there are more seizure-related emergency transports now than ever before. I know: I have friends who are volunteer ambulance drivers. However, no one has done a comprehensive study of seizures that result in death and their link to aspartame, so it's hard to make any assumptions on this matter. However, I am really bothered about the four deaths linked to NutraSweet. Not to mention some pretty creepy research results I've been reading – holes in the brains of laboratory rats, dead foetuses. I asked a friend a few weeks ago what she knew about the deaths related to aspartame, and she gave me an article about one of the victims."

I walked around my desk to my chair and fumbled through my briefcase. "Let me tell you about this article," I said to the class as I leafed through papers. They nodded in silent agreement, and finding the article, I began.

"The article was written by Betty Hailand and is dated December 7, 1987." Referring to the article which laid out the tragic story, I began telling the class about Patty Cain. "Patty Cain was a beautiful girl. She enjoyed a normal and healthy life until she mysteriously dropped dead at age twenty-three. Cause of death: unknown. Patty's mother, Betty Hailand, witnessed the tragedy evolve.

"Patty was Betty's adopted daughter. She was the 'All-American Girl.' Suddenly, out of nowhere, Patty developed eye problems, experiencing blurred vision accompanied by bad headaches. Betty took Patty to have her eyes examined. The doctor found nothing wrong with her eyesight. Patty and Betty were frustrated because they knew something was wrong." I interrupted the story to tell them what Dr. Jim Bowen once wrote about aspartame symptoms and how often they are misdiagnosed. "'Examination by numerous physicians will usually indicate that there is nothing wrong with the patient. He will be diagnosed as a hypochondriac. For those affected, their world crumbles about them mysteriously. Meanwhile, neither they nor their physicians have an inkling of what's going on.' Keep that in mind, ladies and gentlemen," I told the class before I continued Patty's story. "One day after work, Patty returned to her apartment complaining that her vision was intolerably blurred, and she was experiencing unbearable head pain. She progressively grew worse through the night and willingly admitted herself to the hospital emergency room early the next morning. The ER doctor diagnosed Patty with a common case of the flu. He routinely ordered medication for her nausea, immediately prescribed IV's to be administered to her while in the ER for severe dehydration (she required three IV's), and sent her home after they had done all they could for her. She was told to drink plenty of liquids, which she always did – plenty of diet drinks. She went home and drank countless diet colas to sate her dehydration.

"Two days later, Patty's health was returning to normal. Two days after that, Patty was found dead on the floor of her apartment. Apparently, she died around 4:00pm. Her hands were tightly clenched and her tongue sharply bitten. Empty diet drink cans were scattered throughout her apartment.

"There was no official cause for the grand mal seizure that ended Patty's life at age twenty-three, but after researching, Betty was sure she knew what killed her daughter. She maintained Patty died from using products containing aspartame.

"Betty charged that her daughter was addicted to aspartame. Patty incessantly drank no fewer than six diet drinks every day and perpetually added in excess of five packets of chemical sweetener to every glass of iced tea she drank.

"Despite the cause of death being listed as unknown, Betty never stopped believing her daughter's death was connected to her heavy consumption of aspartame. In the memory of her daughter's 'cause of death unknown' as stated on her death certificate, Betty devotedly battled the NutraSweet Company and fought to get information out to the public concerning the dangers connected to the chemical sweetener they marketed.

"The final chapter in this tragic story is one of even more tragedy and death and to it is added violence. Just this year, Betty was found shot to death in her Vista, California home. She was taking a bath when her assailant broke into the bathroom and shot her while she was in the bath-tub. To date, the Los Angeles police have not apprehended her murderer."

The class sat in stunned silence. I told them that, unfortunately, the friend who gave me the article heard of hundreds of sad stories about aspartame consumption all the time.

I stopped and sat down and the entire room became pin-drop quiet. Breaking the heavy silence, a student asked, "Why don't doctors know about aspartame and its harmful effects?"

"That's a tough one. I think that many physicians are frustrated dealing with adults and children with chronic symptoms of degenerative disease. However, with the number of chemical food additives on the market today, no one physician can be expected to keep up with all the isolated chemicals or the combination of chemicals causing diseases," I answered. Pausing, I looked into the faces of the agitated young people and then continued on, "But there's no excuse for not trying. And no excuse for not being informed! My doctor simply wasn't aware."

"But, Professor," a petite brunette student broke in, "is the information available to doctors? I mean, I never heard about this before."

"Well, I feel after my research that information about aspartame and its containment in NutraSweet hasn't been divulged. I'm a professor. My job is to supply information. So, I'm particularly sensitive to the suppression of data. There is definitely a lack of good documentation on the dangers of the aspartame found in NutraSweet. I can't help asking myself why, if there is all this data, so few know about it. However," I added, "this information, though hard to find, is out there. If doctors are unaware or uninterested or just too busy to look into it, then people like you and me are going to have to find it, reveal the facts and bring them to the rest of the public's attention.

"For instance, one article I've found is written by Robert H. Moser, MD, the NutraSweet Company's chief medical adviser. His comments give me much to think about. Doctor Moser wrote in his article, Serendipity Can Be Nudged: 'Medicine is a magic touchstone. It is an Open Sesame.' If you love people, you can put your all into private practice, or work in an AIDS clinic, or labour on behalf of the World Health Organisation in Sudan, Ethiopia, or Southeast Asia, and help sick, starving, frightened people.

"'If you hate people, you can spend your life plugged into a mountain of shining diagnostic equipment or perched beside an electron microscope. Or you can become a cruise ship doctor and play pattycake with rich, blue-haired widows, or you can become a department chairman at Yale, or you can work for the NutraSweet Company.'"

My recitation of Dr. Moser's article elicited looks of surprised displeasure. I asked, "Does this represent how the NutraSweet Company feels about the public? Do doctors feel this way about NutraSweet?"

There were a lot of heads shaking and shoulders shrugging around the room. I continued, "What it really comes down to is that too many doctors are unaware of the dangers connected to aspartame. Mine was. Adults and children are suffering because of it. Patty Crain is dead because of it. I almost lost my thyroid gland because of it. We should, no, we deserve the facts, the truth."

THE FOOD AND DRUG ADMINISTRATION SYMPTOMS FOR ASPARTAME

```
COUNT... %RECORD... %OCCUR.. SYMPTOMS DECODED
1483 ..... 27.35% ..... 18.53% .... Headache
648 ...... 11.95% ..... 8.09% ..... Dizziness or problems with Wance
557 ...... 10.27% ..... 6.96% ..... Change in mood quality or level
520 ...... 9.59% ...... 6.49% ..... Vomiting and nausea
342 ...... 6.30% ...... 4.27% ..... Abdominal pain and cramps
312 ...... 5.75% ...... 3.89% ..... Change in vision
Z59 ...... 4.77% ...... 3.Z3% ..... Seizures and convulsions
241 ..... 4.44% ...... 3.01% ..... Diarrhoea
206 ...... 3.79% ...... 2.57% ..... Memory loss
203 ...... 3.74% ...... 2.53% ..... Fatigue, weakness
201 ...... 3.70% ...... 2.51% .... Other neurological
181 ...... 3.33% ...... 2.26% ..... Sleep Problems
174 ...... 3.20% ...... 2.17% ..... Grand mal
173 ...... 3.19% ...... 2.16% ..... Rash
157 ...... 2.89% ...... 1.96% ..... Hives
151 ...... 2.78% ...... 1.88% ..... Change in sensation (numbness, tingling)
146 ...... 2.69% ...... 1.82% ..... Change in heart rate
132 ...... 2.43% ...... 1.64% .... Itching
101 ...... 1.86% ...... 1.26% .... Change in menstrual pattern
92 ....... 1.69% ...... 1.14% ..... Other
91 ....... 1.67% ...... 1.13% ..... Change in activity level
85 ....... 1.56% ...... 1.06% ..... Difficulty breathing
84 ....... 1.54% ...... 1.04% ..... Local swelling
82 ....... 1.51% ...... 1.02% ..... Other urogenital
81 ....... 1.49% ...... 1.01% .... Other Sensory Changes
77 ....... 1.42% ...... 0.96% .... Other Skin
65 ....... 1.19% ...... 0.81% ..... Other Metabolic
64 ....... 1.18% ...... 0.79% ..... Localised Pain And Tenderness
60 ....... 1.10% ...... 0.74% ..... Change In Body Temperature
```

57. . . . . . . 1.05% . . . . . . 0.71%. . . . . Other gastrointestinal
56. . . . . . . 1.03% . . . . . . 0.69%. . . . . Speech impairment
51. . . . . . . 0.94% . . . . . . 0.63%. . . . . Difficulty swallowing
50. . . . . . . 0.92% . . . . . . 0.62%. . . . . Other muscular-skeletal
48. . . . . . . 0.88% . . . . . . 0.59%. . . . . Chest pain
47. . . . . . . 0.86% . . . . . . 0.58%. . . . . Fainting
43. . . . . . . 0.79% . . . . . . 0.53%. . . . . Joint and bone pain
43. . . . . . . 0.79% . . . . . . 0.53%. . . . . Other cardiovascular
40. . . . . . . 0.73% . . . . . . 0.49%. . . . . Sore throat
38. . . . . . . 0.70% . . . . . . 0.47%. . . . . Other respiratory
30. . . . . . . 0.55% . . . . . . 0.37%. . . . Unspecified
29. . . . . . . 0.53% . . . . . . 0.36%. . . . Change in perspiration pattern
28. . . . . . . 0.51% . . . . . . 0.34%. . . . Oedema
26. . . . . . . 0.4% . . . . 0.32%. . . . Change in taste
26. . . . . . . 0.47% . . . . . . 0.32%. . . . Difficulty with urination
26. . . . . . . 0.4% . . . . 0.32%. . . . Petit mal
Z5. . . . . . . 0.46% . . . . . . 0.31%. . . . Change in hearing
24. . . . . . . 0.44% . . . . . . 0.29%. . . . Change in urge volume
22. . . . . . . 0.40%n . . . . . 0.27%. . . . Noctumal
20. . . . . . . 0.36% . . . . . . 0.24%. . . . Change in appetite
20. . . . . . . 0.36% . . . . . . 0.24%. . . . Change in body weight
19. . . . . . . 0.35% . . . . . . 0.Z3%. . . . Change in thirst or water intake
17 . . . . . . .0.31% . . . . . .0.21% . . . . . .Change in saliva output
17 . . . . . . .0.31% . . . . . .0.21% . . . . . .Unconsciousness and coma
15 . . . . . . .0.27% . . . . . .0.18% . . . . . .Constipation
15 . . . . . . .0.27% . . . . . .0.18% . . . . . .Unsteady gait
14 . . . . . . .0.25% . . . . . .0.17% . . . . . .Abdominal swelling
13 . . . . . . .0.23% . . . . . .0.16% . . . . . .Problems with bleeding
13 . . . . . . .0.23% . . . . . .0.16% . . . . . .Other extreme pain
13 . . . . . . .0.23% . . . . . .0.16% . . . . . .Wheezing
12 . . . . . . .0.22% . . . . . .0.14% . . . . . .Coughing
11 . . . . . . .0.20% . . . . . .0.13% . . . . . .Eye irritation
10 . . . . . . .0.18% . . . . . .0.12% . . . . . .Change in hair or nails
10 . . . . . . .0.18% . . . . . .0.12% . . . . . .Simple partial seizures
9 . . . . . . .0.16% . . . . . .0.11% . . . . . .Changes in skin and nail colouration
8 . . . . . . .0.14% . . . . . .0.09% . . . . . .Excessive phlegm production
7 . . . . . . .0.12% . . . . . .0.08% . . . . . .Muscle tremors
6 . . . . . . .Q11% . . . . . .0.07% . . . . . .Shortness of breath on exertion
6 . . . . . . .0.11% . . . . . .0.07% . . . . . .Blood pressure changes
6 . . . . . . .0.11% . . . . . .0.07% . . . . . .Any lumps present

Sweet Poison

COUNT	%RECORD	%OCCUR	SYMPTOMS DECODED
6	0.11%	0.07%	Blood glucose disorders
6	0.11%	0.07%	Sinus problems
5	0.09%	0.06%	Evidence of blood in stool or vomit
5	0.09%	0.06%	Dysmenorrhea
4	0.07%	0.04%	Dental problems
4	0.07%	0.04%	Death
4	0.07%	0.04%	Hallucinations
4	0.07%	0.04%	Complex partial seizures
3	0.05%	0.03%	Swollen lymph nodes
3	0.05%	0.03%	Other blood and lymphatic
3	0.05%	0.03%	Haematuria
2	0.03%	0.02%	Shortness of breath due to poison
2	0.03%	0.02%	Difficulties with pregnancy
2	0.03%	0.02%	Change in smell
2	0.03%	0.02%	(Children only) Development retardation
2	0.03%	0.028%	Change in breast size or tenderness
1	0.02%	0.01%	Anaemic
1	0.02%	0.01%	Change in sexual function
1	0.02%	0.01%	Conjunctivitis
1	0.02%	0.01%	Eczema
1	0.02%	0.01%	Dilating eyes
1	0.02%	0.01%	Febrile

Distribution based on 5422 records and 8001 occurrences of symptoms.

THE BEST LITTLE SYMPOSIUM IN TEXAS

I loved teaching at the University of North Texas. Every semester, I assigned a formal debate as a term project in each of my classes. My students studied a controversial current event, and then presented the corporate viewpoint versus the environmental viewpoint on the issue.

That semester I chose the topic: "The Safety of Aspartame in Today's Food Industry." Because they knew of my involvement, my students dug in. They uncovered information on aspartame from the 1960s and 1970s onward. They also attacked the FDA's approval process for food additives.

The debates were so spirited, students not even enrolled in my classes began coming to my office requesting they be allowed to sit in, and even when there was no more room, they came asking for information for themselves, friends, and family members regarding aspartame. The word about the classroom debates spread like wildfire. At times, I felt like a nutritional counsellor rather than a professor.

I had always taught my students that there are two sides to every issue. However, my research on aspartame led me to believe that the companies marketing it only wanted one side of the controversy exposed – their corporate side, exclusively reflecting their company's position. My students presented impressive information on both the "corporate" point of view and the public's right to know. Not only was it a good learning exercise for them but it imparted a good dose of reality.

Shortly after the class debated, I received a phone call from the university's director of public affairs. She expressed interest in my debate project and was very complimentary on its academic success.

Sweet Poison

She asked to speak with me in her office. I eagerly set a time to meet with her.

Sitting down in her cosy office sipping tea, we discussed my illness, how I'd "stumbled" onto the real facts about aspartame as I researched my Graves' disease, the assorted information from the Aspartame Consumer Safety Network (ACSN), and my desire to unleash the truth. She shared a bad experience she had using aspartame and asked, "Do you have any information that might help answer some of my own questions about aspartame?"

"As a matter of fact, I do!" I responded immediately.

Then she surprised me with the real reason for our meeting. "The university has offered to sponsor a national symposium on the safety of aspartame. Would you be willing to direct it?"

I was overwhelmed. I had always held great respect for the academic system, but at that moment I was so proud of the University of North Texas that I felt incredibly lucky to be teaching there.

"Well, absolutely!" I agreed, stumbling over my words. "I am somewhat speechless. I am honoured. This is great!" My words ran together due to the excitement building inside of me. It was as if in that moment my personal struggle to heal myself was suddenly transformed into a tangible tool to help others. All my burdens were worth the battle.

Along with pride and excitement came responsibility, planning and a lot of work. The Geography Department went through hell with me that semester. I received countless phone calls, attended on going meetings and had frequent visits from assorted students. Previously quiet halls became noisy wind tunnels blowing turmoil down the dean's corridor. I shipped out hundreds of news releases and publicity letters, overworked the department's secretaries and blew the doors of the department's budget. I gave the department chairman nightmares, I'm sure, but he never complained to me about it. Instead, everyone encouraged and supported my endeavours.

The university graciously supplied all the printing and mailing of invitations, excellent catering, superb facilities and media equipment, advertising designs and programmes. The symposium was to be a pioneering national forum and would, we all hoped, serve as a critical, positive influence in educating the public about the dangers of aspartame.

Spending many hours poring over letters, research papers, documents and case histories, Mary and I finally outlined an impressive list of speakers to contribute their research and case studies on the dangerous side effects of aspartame. We selected experts, scientists and doctors from all over the country who were researching the problem associated with the chemical sweetener. The university, we felt, would be an appropriate setting for those who had not previously been recognised by the FDA nor heard by the public to present their research and findings. This was an opportune time to let the other side speak out. However, I was informed that the corporate side would be given the floor too. Not only that, the university's director of student affairs informed me that I was to handle the arrangements. The first person I thought of was Sue Ross, NutraSweet's public relations person whom I'd met at the conference that Mary Nash Stoddard held.

I closed the cumbersome door to my university office one morning, hoping no one would knock while I made the dreaded phone call. Using the eraser tip of a sharpened pencil, I painstakingly punched the eleven-digit telephone number. I was sick to my stomach by then. I resisted slamming the receiver down only because I had promised to do this.

Was it wise to confront the force responsible for my Graves' disease? Being asked to direct the symposium was a big responsibility and a big honour. I had to make this commitment and stick to it. So, I swivelled skittishly in my chair, my foot tapping nervously on the linoleum tile. The phone rang once.

"Good morning," a prompt voice startled me. "The NutraSweet Company. How may I direct your call?"

Gulping hard, I said, "Sue Ross, please." I tried to steady myself during the short wait before the manager of the public relations for the company came on the line.

Of course, it was difficult because the symposium was designed to inform the campus community of aspartame's negative side effects. However, I could understand that the university felt a sense of obligation to invite speakers from the NutraSweet Company to express the corporate side.

When Sue Ross picked up the line, I was blunt, clarifying that my objective in the symposium was to present research exclusive to the dangers of NutraSweet. However, in order to have a scholarly presentation of data we needed to hear from the other side whether or not I personally agreed with the position expressed. "You may remember me, Ms Ross. We met in Dallas a few months ago. I discussed my Graves' disease at Stoddard's conference held at the

Dallas/Fort Worth International Airport. Aspartame had made me very sick, figuratively and literally."

Ross avoided a confrontation over the pros and cons of aspartame at that moment and said simply that yes, she did remember me. She was most gracious on the phone and accepted my invitation, adding, "I will need to confirm this with my superiors, but I can get back to you first thing in the morning."

"That'll be just fine," I replied, half hoping she'd change her mind by then.

I added, "Please keep in mind I am inviting you to speak, no one else from your company. I carefully selected all the speakers, choosing the ones I felt would solely present their research and not pick fights."

"Certainly," she replied.

I went over the dates and details with Ms. Ross before ending the conversation.

When I hung up the phone, my heart was beating rapidly but I breathed a deep sigh of relief – although it had been difficult, I'd managed to be professional and civil despite all the suffering I had endured.

The symposium date was set for November 8-9, 1991. The speakers at the historic Symposium on the Safety of Aspartame would be two medical doctors, one FAA medical examiner, the co-founders of the Aspartame Consumer Safety Network (ACSN), one doctor of chiropractic, one UNT professor (me), one consumer-advocate lawyer, one journalist, one television news producer, and one representative of the NutraSweet Company. I invited representatives from the federal and Dallas branches of the FDA to discuss NutraSweet's questionable approval process, but they declined after the university refused to pay all their expenses. I was disappointed by this, but there was nothing that could be done. After all, none of the other speakers' expenses were paid by the university.

After weeks of preparation, the day arrived. I looked out over an audience made up of men and women from all over Texas, Oklahoma, and Louisiana. All the students were extremely receptive. I was confident the symposium would be a success.

Dr. Paul Toft opened the first day with high energy and great enthusiasm. He spoke of aspartame's effects on professional athletes. As a chiropractor, Dr. Toft advised athletes and fitness trainers working in the fields of amateur and professional sports. He was especially interested in golfers' reactions to the artificial sweetener, because one of aspartame's side effects causes nerve problems. He explained that if a golfer shook when trying to putt because he or she developed nerve problems due to aspartame consumption, the effect on his or her game would obviously not be good and could prove very costly for players on the professional circuit.

Toft had dedicated countless hours talking with athletes about this potential danger, offering his expertise to young athletes in high schools and colleges. In his speech he told the audience, "Young athletes work their bodies hard while they are still growing. All young athletes must take special precautions to feed, nourish, and rest their bodies so as not to damage proper growth patterns. As they mature, they reach a performance level in proportion to their strength. If they physically break down the building blocks of growth at an early age," he warned, "they may never successfully reach a healthy adult fitness level."

Toft, young and good-looking himself, told his own personal story about his seven-year ordeal with depression, fatigue, back pain, and weight loss caused by using aspartame. "I could not get any relief from these symptoms until I stopped using products with NutraSweet in them," he confessed. "My experience instilled in me a conviction to speak out against this chemical's availability and secrecy."

He nodded as if his words confirmed his own strong beliefs and I felt they indeed did. "I continuously witness the negative reactions aspartame creates in very active users," Toft continued. "Athletes have high energy levels, and their bodies operate at an accelerated rate during workouts. Toxins," his eyes sought those of his audience, "rocket through the body with a higher intensity than when at rest. Therefore, if you work out hard and then consume a diet drink, the adverse effects can impact you more drastically because your body is pumping at a higher level."

Toft gestured angrily with his hands; the gestures of a man fighting a demonic enemy. "Instead of grabbing a diet drink after a workout, I encourage fitness instructors and their students to drink a bottle of water instead. Keep in mind that the body is operating at warp speed during and shortly after a workout. Whatever you put inside your body at this particular time will induce an accelerated reaction, a reaction to anything," Toft stated. It was obvious he wanted those who heard him to know the depth of his feelings.

Dr. Toft concluded his speech by stirring the audience, "If you want to lose weight and gain physical fitness, don't do it by eating diet foods and starving yourself after you work out. Do it by eating correctly and in proper proportions. Accompany a good diet with daily exercise for the rest of your life. Lose weight by eating right and exercising. What a concept! Aspartame can do an intense degree of damage to those who are physically fit."

Jim Turner, JD, flew into Dallas from Washington, DC for a few hours just to speak at the symposium. Turner, an attorney, consumer advocate and author, had joined Ralph Nader in founding the Centre for the Study of Responsive Law in 1969. He supervised research studies on the Food and Drug Administration and the National Institutes of Health. He also participated in regulatory and legislative activities affecting the FDA, the Environmental Protection Agency, the Department of Agriculture, and the National Institute of Mental Health.

"Mr. Turner," I said introducing him, "will you tell those gathered here why you've dedicated years to fighting aspartame?"

"It's a bad thing," he answered, so eloquently and earnestly spoken that his words hung in the large hall. "It's hurting people. This knowledge gives me the energy to keep on it.

"The central point," he went on, "is how do you make society just? If the Food and Drug Administration, with its aspartame approval, violated its mission to protect the public health of this nation, and we can use the law to remove this additive from the market, then it is a positive thing. However, if this doesn't work, then we must be on the alert and find other ways to protect ourselves."

In impassioned tones but with concise reasoning, Turner outlined the history of aspartame approval and tracked the flawed tests and shoddy science. His presentation was eloquent as he talked about the "spin doctors and crisis co-ordinators" that were employed to move the approval process through the FDA.

Admonishing the government and calling for action from the public he stated, "The aspartame experience is a political fix. The FDA approval process doesn't work. The burden of proof has now shifted to consumers to prove aspartame unsafe."

Then, his voice quieting, Turner told of an insensitive remark made by an attorney for the NutraSweet Company about Ms. Shannon Roth, founder of Aspartame Victims and Their Friends organisation. "Roth claimed she was now blinded in one eye due to the methanol within aspartame. She had used NutraSweet products for two years. During that time period, she had experienced constant headaches, dizziness, insomnia, malaise, memory loss, and depression. Every morning, Roth drank four to five cups of coffee with two packets of artificial sweetener in each cup. She also drank diet soda and coffee with sweetener containing aspartame throughout the day, and, in the summer, five to six glasses of iced tea with NutraSweet.

"In 1984, Roth looked in the mirror and noticed a black spot in the centre of her eye. Four days later, she saw only total darkness. She was admitted to the hospital where she underwent numerous X-rays, a spinal tap, and a CAT scan. The diagnosis: retrobulbar optic neuritis, 'blindness in one eye

"Do you know how the attorney addressed her? When the NutraSweet attorney asked me about Ms. Roth, he said, 'How's Blinky?'"

Gasps could be heard from the audience.

Turner continued speaking to the appalled crowd, but turned to the issue of the FDA's 1981 approval of NutraSweet. Turner stated, "When studies are done where the majority of the research is paid for by the industry, it is not scientists' integrity that is the problem. It is not the intention of anyone involved in the process that is the problem. It is the system itself that is the problem. The original issue becomes skewed in a direction where certain questions that would normally be asked by critics are not asked. In fact, most questions raised by critics are never funded for research. It is very important that both those on the side of the additive, and the critics of an additive, have equal resources to do their research."

Turner continued, "This is the only way we can ever sort out whether a product like NutraSweet is effective or not, is safe or not! A substantial proportion of the studies that NutraSweet approval relied on did not appear in peer-reviewed journals and were not peer-reviewed. This was one of the most important weaknesses in the record."

Turner left the crowd with a bitter taste in their mouths. But it was only the beginning. The next speaker was Mary Nash Stoddard. Mary was a charismatic speaker. She was Dallas' first female disc jockey on KVIL radio in the mid 1960s. She went on to produce radio talk shows and commercial videos. In 1977, Mary was sworn in as an appointed judge in Dallas for the State of Texas Board of Adjustment and served on the bench for eight years. She spoke on the economic impact that the use of aspartame has caused in the workplace. Citing a correlation between rising insurance and health costs and aspartame use, Mary told the audience that when employees suffer with headaches, experience mood swings, increase their own and their family's health problems, and accumulate downtime and a loss in productivity, American corporations lose hundreds of thousands of dollars.

As I listened to her speak, I did some calculations of my own. If I had my thyroid removed, it would have cost me in excess of $5,000 in medical bills, plus $90 a month in medication for my entire lifetime. On the other hand, my natural recovery cost me and my insurance company pennies in comparison.

As Mary continued, she shared her personal experience with a life-threatening blood disorder she contracted as the result of using NutraSweet products. "My disease disappeared when I stopped using all products containing aspartame. When I confronted the NutraSweet Company with my accusations, they responded that this type of 'recovery' didn't count because it wasn't a 'controlled laboratory study.'"

Mary oversaw the Aspartame Consumer Safety Network from its inception in 1987. The hotline is available to allow anyone to report a harmful reaction to aspartame.

"Through the years, the stories I have heard and the statistics I have gathered are mind boggling!" Mary has persevered in keeping public the research and information proving aspartame unsafe. Most information available on the dangers of aspartame can be sourced to Mary's many years of ardent efforts to get the truth out. As I listened to her, I scanned the audience and saw rapt faces focused on every word Mary uttered. It made me realise that she is one special lady, and reminded me of a modern day Madame Curie.

In November of 1987, Mary provided vital testimony to the United States Senate in Washington, DC, during the third Senate hearings held regarding the dangers of aspartame. Since then, she has worked with the FAA (Federal Aviation Administration) and international commercial and military pilots regarding dangerous reactions pilots experience in the cockpit while using aspartame. This dangerous aspect of the aspartame issue – its effects on pilots – was what Mary turned to next.

"When pilots encounter changes in air pressure as in takeoffs and landings, blood vessels in their brains dilate and constrict to the point that aspartame's ability to penetrate the blood brain barrier alters normal brain function," she explained. "Pilots are then prone to an increased frequency of seizures. They are more susceptible to flicker vertigo or to flicker-induced epileptic activity after consuming aspartame immediately before and during flight. All pilots are potential victims of sudden memory loss, dizziness and gradual loss of vision during instrument flight when vertigo is most likely to be experienced. They may even get hypoxia at high altitudes as a result of the methanol in aspartame binding with oxygen molecules."

A hush descended on the room. I shuddered to consider how many dieting pilots drank coffee or hot chocolate or chewed gum with aspartame to neutralise pressure build up in their ears during flights.

Mary revealed, "I am hearing from hundreds of pilots who call the ACSN hotline to request information for themselves or someone they fly with, or to share stories of captains and first officers passing out in the cockpit for up to ten minutes after drinking or eating diet products during flight."

Visibly bothered by this information, the audience shifted in their seats. Mary continued on, sharing one of the most disturbing stories of all, that of former Air Force Major Michael Collings. "On October 4, 1985, Air Force Major Michael Collings landed his F-16 at Las Vegas' Nellis Air Force Base after routine manoeuvres," she began. "Two hours later, Collings suffered a grand mal seizure. Major Collings was an experienced fighter pilot. He often flew wing-tip to wing-tip military manoeuvres. Twice, Collings was approached to join the Thunderbirds, the United States Air Force precision flight demonstration team. An avid runner, he stayed physically fit and was always healthy. But," Mary's voice had a razor edge, "Collings' career as a pilot ended that night."

With people on the edge of their seats, engrossed, Mary's voice quickened, "Major Collings traced the patterns of his tremors and seizures directly to aspartame. When he was stationed on remote bases where there were no diet drinks, he had no tremors. When he resumed his use of NutraSweet, his tremors also resumed. For the first time in his flight career, Collings began losing control in the cockpit. He was sent to the flight surgeon who grounded him from flying until his problems disappeared. He didn't know to stop drinking diet drinks and continued suffering vision loss and cerebral complications. The tremors progressively increased in severity from 1983 until 1985, ultimately culminating in the grand mal seizure that put him in the hospital and ended his flight career. He was grounded and assigned to a permanent desk job, and instructed 'not to say anything.'

"After becoming aware of the problems associated with aspartame and particularly those relating to pilots, he stopped using diet products. His health problems which had developed over the past years immediately disappeared. He went directly to the flight surgeon pleading complete recovery, but did not receive the support he expected. He was not reassigned to flight status even though his problems vanished on October 6, 1985, the day he quit using aspartame."

Mary explained that Major Collings testified along with her before the United States Senate's Labor and Human Resources Committee on November 3, 1987. He stated that the tremors and seizures he first experienced in 1983 as a result of drinking three quarts of aspartame-laced powdered fruit drink plus two or more diet sodas a day continued until 1985. While in the hospital after his grand mal seizure in 1985, his father brought him a newspaper article on "NutraSweet and Seizures." Collings said that two days later he quit all NutraSweet, and his seizures and tremors stopped.

"Ladies and gentlemen, I was proud to hear Collings close his congressional testimony by saying, 'My career as a pilot is over. My concern is for others who consume aspartame, especially people who fly fighter-type aircraft and those in high risk jobs.'"

According to Mary's records, many pilots who had lost their medical certifications could trace their health problems to aspartame. She told of her meeting with FAA Chief Deputy Flight Surgeon Dr. Jon Jordan in Washington, DC in December of 1988. "During the meeting, it was decided that the Mot Hotline be established for pilots and flight examiners to confidentially report adverse reactions to aspartame. Air Force, Navy, commercial, and general aviation pilots inundated the hotline with reports of reactions to aspartame that could possibly terminate their flying careers."

The audience burst into applause.

Mary then relayed the nightmare story of another victim. "Tom Clark, an Air Force pilot who flew C-130s, was grounded for minor seizures he blamed on his use of aspartame. When he stopped drinking all diet sodas with NutraSweet, his seizures 'just went away.' Even though Clark's seizures were temporary and he never experienced another seizure after he stopped drinking diet colas, the Air Force would not grant him permission to fly. Clark said, 'There's nothing medically wrong with me. There will always be well-trained pilots to defend this country. To have us tripped up with something as simple as a diet soft drink is unfortunate.'

"And it is not one or two pilots who have had bad experiences with aspartame. In fact, many pilots have told me they avoid all aspartame forty-eight hours before their yearly physicals because they claim it spikes their EKG."

Over horrified gasps, Mary continued. She spoke of Phil Moskal, Ph.D., who performed a study of pilots using aspartame in 1990. Dr. Moskal, a professor of microbiology, biochemistry, and pathology and the director of Public Health Laboratories, was also an instructor-pilot.

Citing his own research, Dr. Moskal warned that, above 29,000 feet, "things begin to happen."

"I have already shared some pilots' stories that reveal adverse side effects pilots experience," Mary said. "Let me share another with you." Mary read a transcript from a telephone call she received on the Pilot Hotline.

"'Aspartame Consumer Safety Hotline, how can I help you?'" she began and continued reading. "'Someone put this article on the bulletin board in our pilots' lounge. It has your number on it. I want more information, but I also wanted to tell you what happened to me. I'm a pilot with [a major airline]. Last year, I experienced a grand mal seizure before a trip.'" Everyone in the room sat up, listening with rapt attention. "'Several weeks before that, my jaw started twitching. I was drinking a diet shake two times daily to lose weight. I had no medical problems before that. My doctor suggested it might be due to the aspartame in the product, but we couldn't call it that because the FAA does not recognise aspartame reactions in their regulations. At my doctor's suggestion, I quit the product and have had no problems since.

"'I lost my medical, but my doctor figured that if we could call my seizure an alcohol-related problem, I could go through an alcohol rehabilitation clinic and that might do it. Turns out it did. I've never had a problem with alcohol, but we had to go through the motions so that I could keep my job. The doctor even used some fancy syndrome name and it worked.'

"Pilots' reactions to aspartame need to be thoroughly investigated," concluded Stoddard. "Much research has yet to be done, but, considering the potential risk to other people's lives, it is wise to err on the side of caution when it comes to 'drinking while flying.'"

Mary stepped down from the podium as the audience rose to applaud her. Her comments on pilots and aspartame seemed to dramatically affect the audience, so I suspected the next speaker would knock their socks off.

FIT TO FLY?

The next speaker was Dr. James B. Hays, Designated Medical Examiner for the Federal Aviation Administration (FAA). Dr. Hays performs medical examinations on pilots that are required for medical certification to fly.

Dr. Hays stood at the podium with a burly Texas stance of authority, and in his Southern drawl described some of the more dramatic cases of pilots stricken with side effects of sweet poison. Dr. Hays identifies pilots who have had negative reactions to aspartame. He has recorded numerous case histories over the years and submitted a number of articles to medical journals concerning the matter. He currently sees one to two people every month who have a problem caused by aspartame. His obvious commitment to the safety of the airlines and the thousands of passengers at risk daily was evident.

Dr. Hays aids pilots who have had seizures in flight and documents their reactions to aspartame over time. If pilots have a medical problem affecting their ability to fly in any way, they risk losing their medical certificate to fly. Pilots take this very seriously because their careers are at stake.

Dr. Hays shared the story of C.K., a pilot who suffer a reaction to aspartame, costing him his medical certificate. Unlike the military stunt pilot Michael Collings, C.K. fought the FAA for several years until he succeeded in reinstating his medical clearance. According to Dr. Hays, C.K. had a five-year history of drinking more than two litres of diet drinks daily and was adding large quantities of aspartame to iced tea. His head deteriorated slowly over a five-year period, beginning with malaise, a feeling of tiredness and ill-being.

"C.K.'s symptoms progressively worsened," Hays related. "He began to

93

get headaches which became quite severe. He started experiencing a lack of recall rather than actual memory loss. As his symptoms magnified, he began feeling a loss of control referred to as a syncopic pass. C.K. experienced a syncopal episode in July 1982 at which time he had numerous feelings of wanting to 'pass out.'"

Hays' voice rose, "C.K.'s symptoms continued until he was eventually hospitalised. A complete neurologic work-up was performed," Hays went on, "including a negative EEG and CT scan. He improved somewhat initially, but over a period of time returned to his normal consumption of NutraSweet. His symptoms returned and progressively worsened, actually becoming quite severe. In the spring of 1986, he had what I describe as a 'seizure.' He bit his tongue, but was not incontinent. C.K. stated that he again had multiple feelings of almost 'going out.'

"Following the seizure," Hays continued, "he developed a severe headache and blurring of vision. His symptom improved over the following two weeks. Two weeks later, upon reinstating his use of NutraSweet, his symptoms returned.

"At this time, C.K. put two and two together, identifying his symptoms with his use of NutraSweet/Equal. He immediately stopped using all aspartame. His symptoms instantly improved. His feeling of 'going out' disappeared, followed by fewer headaches and memory loss episodes. His family confirmed he was less irritable. Within six weeks of discontinuing the drug, and yes," Hays paused, his concentration intense as he stared at the audience. "I refer to aspartame as a drug," he paused again. "C.K.'s headaches were infrequent and mild. His memory loss had disappeared. His vision remained impaired to a minimal degree for several months. As long as nine months later, he noticed that he no longer required large amounts of antacids. During the course of his withdrawal from aspartame, if he sampled any NutraSweet or Equal, his former symptoms would immediately return. In 1990, two years later, C.K. was entirely symptom-free. Presently, he refuses to participate in any drug studies which might require him to ever ingest aspartame."

Hays said that C.K. fought the FAA until he succeeded in reinstating his medical certification. C.K. was convinced that aspartame was dangerous for him and for all pilots. He believed that under the workload of a single-piloted aircraft operating in instrument conditions, partial impairment of capability was just as fatal as total impairment or unconsciousness.

"I wholeheartedly agree," Hays confirmed, his voice showing his genuine concern.

In his practice, Dr. Hays has observed numerous aspartame side effects in several pilots over the years. His professional observations prove that symptoms are dose related. "More aspartame use over a longer period of time increases the probability of headaches," Dr. Hays said during his speech.

Dr. Hays went on to talk about a study he conducted with ten subject patients. "The patients' aspartame symptoms were notable enough to record, and they occurred in a patterned series," he stated.

Dr. Hays strongly suggested that in a work-up for headaches, an aspartame history should be taken. He recommended the patient be placed on a clinical test:

• Increase the dose of aspartame for one to two weeks to note symptoms, if they are unclear in the beginning

• Record each symptom in a diary.

• Discontinue aspartame for one to two weeks.

• If symptoms clear up, confirm the diagnosis by returning to the previous dose of aspartame for one to two additional weeks.

• "After using this diagnostic approach," Dr. Hays told the audience, "none of my test subjects lasted more than two weeks before all their aspartame symptoms returned."

Dr. Hays concluded with a dire warning that aspartame, "is a significant hazard to transportation safety." He stated, "Many physicians in the Federal Aviation Administration (FAA) are aware of the aspartame situation, but are bureaucratically linked to the FDA, which is in charge. The FDA refused to admit that they had made a mistake in approving aspartame, and the FAA had to abide with the FDA as the FDA is the drug or 'additive' approving authority."

He proposed that a simple clinical test like the one he had earlier mentioned be implemented to establish if aspartame is indeed the cause of suspected medical symptoms. "Clinical research needs to include human subjects," he stated. "And, by using subjects shown to be previously affected by aspartame, all subjects should have a sufficient dosage of aspartame over a several-month time period."

He ended his presentation with a statement I will never forget. Standing forthright and trustworthy, his eyes and voice showing his deep sincerity, "When a DC10 slams its nose into the ground, I will be the first one to stand up and say, 'I told you so!' I'll repeat those four words: I Told You So!"

Sweet Poison

The room was silent. The audience was overwhelmed learning about dangers in the air they'd never heard before. I found myself gritting my teeth in silence waiting for the next speaker, Barbara Mullarkey, to continue our enlightenment when she began her speech.

Barbara Mullarkey was a determined journalist who began her investigative articles regarding the safety of NutraSweet in the early 1980s. Mullarkey told the audience of the many "Letters to the Editor" her newspaper received from the NutraSweet Company's medical spokesperson, Dr. Robert Moser. A nutrition-ecology columnist for the Wednesday Journal in Oak Park, Illinois, Barbara continuously investigated the effects toxic foods and environmental pollutants had on people and the environment. She was accustomed to opposition like Dr. Moser's, but "You really get tired of them," she admitted to the audience.

Mullarkey told how her investigations of NutraSweet started when she called the Chicago office of the FDA routinely asking, "What's new?"

The FDA consumer affairs officer replied, "Aspartame!"

Mullarkey asked, "Aspar-what?"

Intuitively, Mullarkey admitted that she felt aspartame would be a problem from the beginning.

"Something wasn't right about the way aspartame was approved," she said. "This signalled caution in my mind."

When she learned from independent studies of the many harmful results, versus the company's own research results of product safety, she knew something was suspicious.

She still posed the same question after writing more than 35 articles on "aspar-what". Many of her articles featured the background of the artificial sweetener, and during her investigations she established many contacts among the media and the government supporting her mistrust.

"What can we do about aspartame?" she asked the crowd, her words ringing out. "Eat food!" she told them with conviction. "Most of what people eat from supermarket shelves, fast food outlets and restaurants are laboratory concocted factory foods.

"A friend once told me," Mullarkey confided, "if you know the truth and don't speak it, you are a liar." So, Mullarkey told the audience, she was fuelled by her conviction to do, say, and write what is right. The audience obviously appreciated her convictions – their applause was thunderous.

By now the atmosphere was so charged with electricity and righteous anger that I felt appreciative when H. J. Roberts, MD, opened his presentation with a joke. With a little levity he helped us all remember to have a sense of humour about the goals we were trying to achieve and impart. Dr. Roberts, an internist, endocrinologist, internationally known medical consultant, researcher, and author of more than 200 original articles and letters and five acclaimed texts, believes strongly in the healing power of laughter.

"The history of aspartame and its approval and support from the FDA are a farce," Dr. Roberts told the crowd. "The lack of information warning of the medical danger is very real, planned and misleading." Standing at the podium, Dr. Roberts, a well-respected physician, illustrated through his vignettes just how insidious this matter had become.

"This 1987 cartoon published in the Medical Tribune," opened Roberts, "illustrates a chapter in a book I'm working on about the confusion and memory loss associated with aspartame use. It depicts a doctor sitting at his desk holding his hand over the phone receiver, yelling across the room to his receptionist, 'Ms. Ames, what's the name of the lady with the memory loss problem?'"

Roberts continued over the laughter. "I believe 'this problem' is widespread, yet remains shrouded in secrecy. Many regard me as an aspartamologist because of my extensive pioneering studies in this field of medical concern. Then, there are those who consider me 'Corporate Enemy Number One.'

"Aspartame is 'the tip of the public health iceberg,'" he asserted.

Like Mary Stoddard, Dr. Roberts had received hundreds of letters from people afflicted with 'aspartame disease' searching for answers that traditional medicine could not supply. He cited the fact that 85% of the complaints received by the FDA pertaining to foods and food additives concerned aspartame products. He explained, "As of this year, almost six thousand complaints against aspartame have registered over eight thousand various health symptoms. The most common complaints include severe headaches, confusion, memory loss, depression, convulsions, dizziness, insomnia, impaired vision, nausea, diarrhoea, rashes, and weight gain."

As I stood offstage listening, I couldn't help but think the symptoms he just described sounded awfully familiar.

Dr. Roberts made the point succinctly, "My patients with low blood sugar, known as hypoglycaemia, and with diabetes are adversely affected by 'diet' products." I was sure that Roberts was right about this because when my diabetic sister, Beth, stopped using aspartame, she decreased her daily insulin intake, and my hypoglycaemia completely disappeared when I stopped using NutraSweet. Doctor Roberts is on to something, I thought.

Roberts described many case histories of diseases where a patient's adverse reactions pointed directly to aspartame. Based on his findings, he theorised, "The cause of diseases such as Professor Hull's 'Graves' disease' is indeed sourced to aspartame's toxicity."

I felt like a heavy weight was suddenly lifted from my shoulders. Dr. Roberts just announced to the world out loud what I had learned in my difficult trial. He confirmed I was right about aspartame causing my 'Graves' disease.' How long had I waited for this? It seemed like forever, but I was happy to hear those words from the mouth of a physician, so the wait seemed unimportant.

Dr. Roberts continued speaking, moving to the pilot issue. Like some of the others, Dr. Roberts worked with pilots who experienced dangerous reactions to aspartame while in flight.

"Most doctors are not well informed concerning the dangers of aspartame, especially pertaining to pilots," he stated. "Coupled with misdiagnosis, pilots have a lot to lose. Pilots are likely to pass their symptoms off as temporary, or caused by something else. They commonly won't mention anything to their doctors, particularly during an FAA exam; therefore, they don't receive the help they really need. It may negatively affect their medical certification. Too many pilot cases remain hidden."

To illustrate the results of this phenomenon, Dr. Roberts spoke of an article he wrote for the Palm Beach Post, on October 14, 1989, concerning the growing number of reports linking aspartame to pilot and driver error. He told the audience, "Recent plane accidents warrant further inquiry into pilot error caused by aspartame. I'd like to raise the question whether the co-pilot who inadvertently hit the disengage button before the nineteen 1989 US Air crash had been drinking diet while in the cockpit. Was it ever investigated?"

Roberts went on to describe a study he published on 157 people who experienced aspartame-induced confusion and memory loss. The subjects included pilots who developed neurologic-psychiatric symptoms, including convulsions and visual problems. He noted the pilots in the study were all health-conscious professionals who used aspartame in an attempt to avoid sugar.

Roberts shared another 1989 article. This one was from the Australian

Aviation Safety Digest published on two pilots' stories of reactions to aspartame while in flight. "Neither pilot knew what caused his flight reactions," he said. "But after removing aspartame from their diets, each pilot was able to identify, and report, the correlation.

"Preceding a night flight, one Australian pilot drank two cups of artificially sweetened hot chocolate," he continued. "During the final leg of the flight, the pilot was unable to see the instruments clearly because of blurred vision. The pilot said, 'I remember the controller asking me my airspeed. I was confused and unable to read or interpret the instrument. I gave him my DME (Distance Measuring Equipment) digital readout which was in large, bold numbers. I maintained altitude by keeping the big white needle straight up and down on the altimeter. I felt apprehensive, insecure, and way behind the airplane. I knew if I had a real in-flight emergency, I would be unable to handle it since I was already in an overload condition.'"

Roberts cited the work of Dr. Richard Wurtman, a neuroscientist at MIT, who stated that consuming large quantities of aspartame created a chemical imbalance that lowered the threshold at which many people have epileptic seizures. He also mentioned research from Dr. Woodrow Monte of Arizona State University and Dr. Morgan Raiford of Emory University who both found links between the methanol in aspartame and optic nerve damage. They stated that you could get the same amount of methyl alcohol in a litre of diet soda as in four-fifths of the alcoholic beverage, Wild Turkey!

"Retesting aspartame as an ulcer drug may be the solution for pilots who have suffered negative reactions to NutraSweet," stated Roberts. "Drugs are required to be tested far more stringently than food additives. To prove whether a new drug is safe, it must be tested on humans over a long period of time under strict surveillance."

Roberts pointed out that many drugs affect pilots' brain chemistry, creating imbalances during flight. However, food is not recognised as a cause for seizures. Yet, Dr. Roberts argued, "Did Major Michael Collings and C.K. have seizures sourced to 'food' or to a 'drug?' I believe they experienced serious neurological reactions to aspartame. Not food reactions. Aspartame is not food."

Roberts stated that pilots know that drugs and alcohol don't mix with flying, and that over-the-counter drugs and prescription drugs had also been proven to negatively affect pilots during flight. He explained that there are different classifications for the medical use of drugs in flight, ranging from permitting flight duties while taking a drug to preventing flying entirely.

However, since aspartame was approved as safe by the FDA in 1981, he told the crowd, many pilots had not linked its use to serious health problems. "But it is a serious problem," confirmed Roberts.

Then Roberts did an insightful little exercise to illustrate how detrimental this type of health problem could be for a pilot. Specifically, Roberts outlined a pilot's medical requirements to see just how threatening aspartame could be:

"To exercise the privileges of pilot-in-command, all pilots have to carry a valid medical certification of an approved class. This secures that at least once during the required time period, pilots have been physically checked by an authorised physician and deemed safe to pilot an aircraft. The main concern is directed toward conditions that suddenly become incapacitating such as a heart attack, seizures, vertigo, etc., all of which might lead to the loss of aircraft control and probable accident. Pilots who have a history of heart problems, diabetes, kidney stones, fainting spells and the like might not be able to renew their medical certification, ultimately costing their careers in some cases. Once a pilot has lost his or her medical clearance, it is extremely difficult to reinstate, even if the condition has been corrected. General good health and vision are necessary for medical certification regardless of other criteria. So, pilots have to pay special attention to their health. FAA physicals are thorough, sometimes more so than a regular physical exam.

"Federal Aviation regulations parts 61, 91, 135, and 121 pay special attention to the requirements for the use of alcohol and drugs, the carrying of certain substances, and the submission to drug and alcohol tests for criminal reasons, as well as for health and safety in flight. Pilots live by these rules."

"But what about prescription and over-the-counter drugs and food additives?" Roberts went on. "Many drugs can have a serious effect on pilots performing duties as pilot-in command such as Major Collings. Consider the reactions that may be caused by the combination of drugs and food additives."

But, Roberts explained, most pilots consider food ingredients as safe. He gave the example of a pilot who has taken an anti-histamine and is chewing sugar-free gum with NutraSweet, plus drinking hot coffee with two packets of Equal in it, while in the cockpit. Roberts offered the following food for thought: The guy is simply watching his weight and has a blasting sinus headache. But you're his passenger…"

Eyes widened and heads shook throughout the auditorium.

Roberts cited Medication and Flying: 'A Pilot's Guide' by Dr. Stanley R. Mohler as a good reference for pilots concerning drugs and flying. Dr. Mohler wrote, "A Pandora's box of prescription and over-the-counter drugs can give a pilot a host of 'bad vibes' during flight."

The first two chapters of Dr. Mohler's book focus on the effects of tobacco and alcohol. In it he states that nicotine, carbon monoxide, formaldehyde and hydrogen cyanide are physiologically addictive and produce the following effects: fatigue, headache, chronic cough, shortness of breath – all possibly life-threatening at high altitudes.

Roberts continued to refer to Mohler's work, which included a discussion about over-the-counter drugs. Mohler wrote that Actifed hay fever medicine could leave a pilot dizzy, sleepy, uncoordinated, and could plug up the sinuses. Benadryl, Benahist, Dihyrex, Premodril and other anti-histamines could also cause drowsiness, reduced mental alertness, and blurred vision in pilots during flight.

"Dr. Mohler listed some side effects from using common drugs while in flight. These include ibuprofen, which causes dizziness, skin rash, heartburn and blurred vision; codeine, which causes dizziness, constipation, mental confusion, nausea and potential for drug use; Nembutal, which causes drowsiness, nausea, hangover-like symptoms and reduced mental function; and finally, Valium, which causes drowsiness, fatigue, confusion and headache.

"In combining these effects with the effects of aspartame," Dr. Roberts continued, "pilots could be taking a risk with their flying futures. Depending on a pilot's physical status and tolerance level the immediate effects of aspartame can be severe."

Dr. Roberts wrapped up his enlightening speech by proposing new labelling of aspartame products: "They should say: 'Warning. Use caution if driving a vehicle or flying a plane.'"

The audience cheered this proposal and Roberts' entire speech. I was cheering too. I believed strongly in his pioneering work and conclusions and it seemed like the whole room felt the same way.

There were only two speakers left: the NutraSweet representative and myself. However, there was another person that I invited who enriched the symposium just by being present.

Gailon Totheroh joined the symposium and wore many hats. Gailon is one of the producers for Christian Broadcast Network (CBN) News in Virginia Beach, Virginia. I asked Gailon to speak at the symposium regarding the many interactions he had with the NutraSweet Company and about the stories CBN aired on this controversial subject.

Having previously dealt with Gailon, I felt certain that he would be very forthright. In fact, CBN News featured many stories investigating aspartame safety, and despite numerous threats against airing those stories, the network continued to support the producers who reported the truth concerning this issue.

Gailon once told me about the first threat against CBN. Several years earlier, Pat Robertson, CBN's founder and CEO of the Family Channel, was hosting Mary Stoddard and the late Betty Hailand (Patty Crain's mother) on the 700 Club. Before Mary and Betty appeared on the show, the NutraSweet Company flew representatives to Pat's office in Virginia Beach three times to "talk" him out of airing the clip but he persevered even though three days after the show aired, a mail bomb exploded in Pat's office. Pat escaped injury, but his security guard was seriously hurt in the explosion.

Gailon didn't speak at the symposium, but he did bring a film crew to the university for a week of filming and interviewing. He explained to me that after every story he has produced thus far concerning aspartame and its containment in NutraSweet products, he, his producers, and Pat Robertson had received letters, wires, and phone calls from some employees of the NutraSweet Company attacking Gailon's character, his job performance, and his media future. He told me he didn't want to risk jeopardising the symposium in any way, so he elected not to speak.

Turner, Totheroh, Stoddard, and Mullarkey all spoke of character attacks, threats, and bullying tactics aimed at them whenever they spoke out against NutraSweet products. However, I was still shocked.

STARR WARS

So far it had been a dynamic symposium, the speakers, impressive, the evidence they gave on aspartame, dramatic. As word of the symposium spread, the news hit the Associated Press wire, and the media strongly responded. Reporters from around the world were interested in the symposium, the aspartame issue, and in the American FDA approval process. On this, the second day, the presence of the press added to my nervousness when I started my speech. However, by this point I was relieved to have them there documenting everything that occurred.

I'm not a medical doctor. I'm not an accomplished lawyer or a public speaker. I am merely a victim. Nevertheless, I knew my experience was real and horrible. It was important for others to hear. What I had to say was that I was poisoned by aspartame and that when I had eliminated all aspartame from my diet, I completely recovered from Graves' disease. I mentally psyched myself up on the way to the podium, telling myself, I must alert others to the same danger so they won't fall into the same trap! I opened with my Graves' diagnosis and shared my self-imposed path to recovery.

"I'm angry about the silencing of information about aspartame," I went on. "I am a consumer and deserve to know the truth – that aspartame can harm my health."

I told the audience about something I had picked up along the way. "One day I heard nutritionist and public speaker Ted Broer refer to the 'Ten Foods Never to Eat', calling the top four foods on the list 'The Kevorkian Four.' Aspartame topped his list."

"I have questions, not answers," I said to the attentive audience. "Do we excuse food additives as a cause of illness' "Are the four deaths filed with the FDA no big deal? How many more deaths have gone unknown? Patty Crain's did. Do all pilots know they can have a dangerous reaction to aspartame while in flight? The FAA isn't posting warnings about this. Why not? Some pilots are being told not to say anything. Why?" I asked angrily.

"What unknown present and future dangers are we exposing our children to? How many people suffer headaches, backaches, tension and stiffness, PMS, cramping, bloating, sinus problems, colds, allergies, weight loss, weight gain, hyperactivity, insomnia, depression and bed-wetting while using products with aspartame? These questions have gone unanswered for far too long. I demand explanations."

I was surprised to be interrupted by applause and cheers. Feeling more confident, I went on.

"When I stopped using aspartame," I explained, "my thyroid returned to normal within thirty days. All my mood swings, migraine headaches, PMS and menstrual problems, skin problems and weight problems disappeared within weeks. Many people are currently developing similar symptoms. At younger and younger ages too. With the number of chemicals in our food supply, over time it will become more difficult to identify which chemical is causing what disease. There are too many chemicals to choose from.

"To me," I said with compassion, "one of the most unforgivable aspects of the aspartame controversy is the lack of truth and information. I was completely in the dark."

Thinking of my three boys, I stated, "Children are being raised in such polluted environments today. Not only is the drinking water contaminated and the air quality compromised, but children are eating far too many chemicals. Some babies are fed chemicals in the womb. It's hard to passively stand by and watch a pregnant mother drinking a diet drink. The foetus receives all nourishment before the mother does. Whatever that woman eats or drinks pipes directly to her baby.

"I'm sorry," I said, my voice rising with my impassioned feelings. "But a developing foetus doesn't need a diet soda. I don't think that mother would give her baby a diet drink through a baby bottle. Yet, there is no difference!"

I related something that Mary Stoddard had told me recently. The League, a renowned international organisation aiding nursing mothers, had contacted her asking if there was any research available

104

correlating aspartame to the inability to suckle. They reported receiving over 400 calls a month from mothers with babies unable to suck. "As a matter of fact," Mary had responded, "there is independent research showing animals given aspartame in the womb as unable to suckle, partially due to cleft pallet deformation."

"Epstein Barr, chronic fatigue syndrome, post-polio syndrome, Lyme disease, carpal tunnel syndrome, Muniere's disease, Alzheimer's, attention deficit disorder, Graves' disease, Lou Gehrig's disease, epilepsy, anxiety, phobia disorders, pre-menstrual syndrome, multiple sclerosis, lupus, diabetes, fibromyalgia and eosinophilia myalgia syndrome are only a few of the contemporary diseases being treated with pharmaceutical prescriptions today," I continued with my speech. "With a list like this, any chemical food manufacturer who tries to convince me that their man-made, chemical food additives are no threat to my health is not sincere. I don't buy it."

I wanted so badly to reach out to them to impart the difficult lessons I had learned. "You know, I am weary of hearing the NutraSweet Company profess time and again that their product is perfectly safe and has no connection to any of the symptoms or illnesses I just listed. 'You can't prove it,' their representatives continuously profess. 'Our research proves that NutraSweet is perfectly safe for public consumption.' Their representatives boast that since the 'FDA approved aspartame, it has to be safe.'

"I question this. The tobacco industry continued to deny physical harm from cigarettes based on their research, too. In fact, the FDA approved cigarettes! And cigarettes are still for sale; the dangers are still being debated. A friend once showed me a magazine advertisement from the nineteen fifties they had found boasting that 'more doctors smoke Camels than any other cigarette.' Today, we know better. I am certain that in the future, people will see our advertisements pushing aspartame-laden products and use them as examples of how naive we were.

"Aspartame research has been performed by independent researchers like Jeffrey Bada, Ph.D., Ralph Dawson, Ph.D., Louis Elsas, MD, Reuben Matalon, MD, Woodrow Monte, MD, John Olney, MD, Ph.D., William Pizzi, Ph.D., Diana Dow-Edwards, Ph.D., and Ann Reynolds, Ph.D. These men and women are among the leading researchers with laboratory results of holes in animal brains, dead foetuses, cancer polyps, birth defects and mental retardation in mice resulting from aspartame. Why haven't these research results been made public? Dr.

Sweet Poison

Moser of the NutraSweet Company once answered this question during a radio interview. 'The reason,' he boldly announced, was that 'no one really cares.'

"His statement shocked and enraged me at the time. Now I am less gullible and I see that Dr. Moser knows of what he speaks. If the NutraSweet Company doesn't care, that is their concern. However, complacency on this issue is not acceptable to me.

"I believe merely one complaint, illness, death, or negative reaction of any kind concerning aspartame constitutes a public summons for help. Six complaints registered against saccharin earmarked it unsafe. There have been thousands of complaints against aspartame. Thousands of public summonses for help. This warrants an investigation, a recall, a warning label, or some type of action to be taken.

"Aspartame supporters remind consumers that laboratory animals developed bladder cancer from saccharin. But I want you to remember laboratory animals formed holes in their brains from aspartame.

"This makes no sense to me," I said, cracking a joke to relieve the tension. "Maybe something obvious has slipped through a hole in my brain." The audience chuckled, using the levity as an excuse to shift around a bit. "I feel like I'm witnessing a crime but am mute when I call out for help. I felt this way in the hospital when I was diagnosed with Graves' disease. And I feel this way now," I confessed to the audience.

"A few years ago," I said as I looked at Paul Toft, "Doctor Paul Toft wrote a personal letter to a syndicated morning talk show hostess requesting she air a segment on aspartame and NutraSweet products. She personally replied to him in a letter, saying her 'show is very interested in this issue, but is not willing to exclusively air the story.' She wrote, 'Only... if other major network airs a segment on NutraSweet at the same time... will my producers be interested in considering airing the story.'"

Shifting my gaze to Mary Stoddard, I continued, "Prior to this, a well-known prime time network news magazine flew one of their producers to the ACSN home office in Dallas to photocopy records of research and case histories in preparation to air a story on NutraSweet. Before the segment was slotted to air, Mary Stoddard received a phone call from the show's producer informing her that their advertising department warned them 'not to air the segment or they would pull the show.'

"The fear these networks demonstrated is really unfounded. To date, no legal action has ever been taken by the NutraSweet Company against anyone in the United States who has spoken out against aspartame," I revealed.

"My point is this: Aspartame is dangerous and unhealthy and anyone who knows this should not be afraid to speak out against this product and those who market it. Only when the public has been made aware of this health risk can we look forward to change. Thank you."

The final presentation was saved for Sue Ross, the manager of public relations for the NutraSweet Company. I didn't place her last because she was from NutraSweet. I slotted her as the final speaker because it was the logical line-up since she would be rebutting the point of view of those who spoke before her. I think she was bothered by this, but I couldn't be sure.

When I introduced her to the audience, she smiled at me as she walked up to the podium. Everyone was anxious to hear her comments, including me. I expected them to be similar to the ones she made when we previously met. I was shocked at what I heard! Ross seemed to me very tense, defensive, and emotional.

She began, "When it comes to presenting an open-minded scholarly debate, I'm afraid the organisers of this symposium have missed the mark by a wide margin."

"We're here to protest the way this symposium has been organised," she continued, "because we believe the university and its students have been misled."

My mouth dropped wide open. This isn't a debate, I angrily thought. I knew I made that clear to her when I called her.

As I tried to calm myself, I began to think reasonably about the whole thing. I realised she couldn't say much else. She couldn't say, 'You guys are right. My company is awful. It has deceived the American public all along.'

She obviously believed in her company, and despite my disagreement with that belief, I felt she was entitled to her point of view.

Ross continued to defend NutraSweet announcing that she was a user and was pregnant. She praised the accomplishments of her company and Robert Moser, MD. Even with my calm realisation about Ms. Ross' difficult position, I couldn't help my mind from racing backward, and my thoughts from returning to my hospitalisation and to my doctor's warnings that I would die if I didn't destroy my thyroid gland. I re-evaluated my "miraculous" recovery and revisited my doubts and fears. My experiences were proof of the seriousness of this issue.

After a little less than an hour but what felt like an eternity to me, Ross stepped down from the podium, refusing to answer any questions. She dramatically placed a large stack of business cards before her and invited anyone who wished to speak with her to call the phone number on the card.

Followed by a line of war-like students and frustrated reporters dashing closely behind her, Ross suddenly disappeared, just like the first time we met. It was a dramatic ending to a gripping symposium. After the room cleared, I took time to reflect on what I learned from the symposium and during the weeks of preparation. I felt I had a better understanding of diminishing consumer rights and freedom of speech, and I held a renewed respect for the relevance of acadaemia. I was most disappointed in the FDA. I was particularly disillusioned to learn of the influences major corporations have over the media and within research laboratories, not to mention on the general public. Most important of all, I learned that concerned people could make a difference. No textbook would ever teach me what the symposium did. One reason I enjoy teaching college so much is I really admire the courage young students possess. Not merely my students, but all students. They have faith in ideals and follow their dreams. Students believe they can change the world for the better. The symposium gave them the means.

The entire campus rallied against aspartame shortly after the symposium. They were up in arms at the NutraSweet Company for not alerting the public to the dangers of the chemical sweetener they marketed so successfully. The students proposed all aspartame be banned from campus. They presented a resolution to the student council requesting:

• The Student Association fully support a petition calling for the University of North Texas administration, food services, and all other applicable UNT agencies and departments to discontinue the purchase and/or acquisition of any and all products containing aspartame.

• To remove from all dispensers, counters, machines, etc. those products containing aspartame.

• To refuse permission for the distribution in any manner of aspartame products on the University of North Texas campus.

The Student Association spent more than eight weeks studying the legislation. They voted to educate all students on the dangers of aspartame and to write to the government about the dangers of the chemical sweetener. The Student Council Bill Number F91-14 was amended to include: "On a monthly basis, the Student Association will sponsor a table with information concerning aspartame, as well as information about other artificial sweeteners."

The students were not the only ones calling attention to the symposium and the dangers of aspartame. Barbara Mullarkey published an article recounting the symposium and the students' responses. Her article elicited a reply from the NutraSweet Company.

"Scenes from the Best Little Aspartame Symposium in Texas" From the Wednesday Journal of November 20, 1991 by Barbara Mullarkey writing from Denton, Texas.

"An historic symposium on aspartame/ NutraSweet Equal happened November 8-9 at the University of North Texas (UNT) here. It was an honour for me to present a journalist's overview of this synthetic sweetener. On November 13, the UNT Student Association voted to sponsor:

"• a letter to the FDA calling for a review of independent studies concerning the safety of aspartame, and

"• a three-day table in the (student) union with information pertaining to aspartame and/or other artificial sweeteners... once a month, every month during the Spring '92 semester.

"• The Student Association felt it 'reasonable to question the safety of aspartame due to their concern for student safety.' Important roles were the 'personal testimonies indicating a relationship between ingestion of aspartame-containing products and various mental disorders.' The Student Association felt it 'reasonable to question the impartiality of the FDA's decision to approve aspartame as a food additive.'

"Aspartame seems to control the information and airwaves through its slick, incessant ads. Students have their own agendas. Hooray for students' examples!"

In response, Dr. Moser of the NutraSweet Company wrote the following letter, "Once Again, Mullarkey and NutraSweet at Odds" to the Wednesday Journal of December 11, 1991. It was signed Robert H. Moser, M.D. Chama, New Mexico:

"Once again, I found myself obliged to respond to an outrageous column by Barbara Mullarkey. She reported (in her November 20 column) on a two-day aspartame symposium arranged by a geography teacher at the University of North Texas, who attempted to gather all the known individuals hostile to NutraSweet under the same roof. She did assemble enough for eight hours of harangue; it was replete with the usual barrage of anecdote, innuendo and opinion familiar to those who read Mullarkey. Not one shred of scientific evidence was presented. I was purposely not invited. The NutraSweet Company did send Sue Ross, manager of public relations, who expressed unhappiness for this lopsided, "stacked-deck" affair. Hardly in the great American tradition of balanced, town-meeting style of event, Sue had fifty minutes...

"Shame on you, Barbara. The students at North Texas and the readers of Wednesday Journal deserve better."

I felt the real shame fell on the company which continued to market a product with such potentially damaging side effects on the public's health. But company damage control continued. Another response of the NutraSweet Company to Mullarkey's article was the following letter from Sue Ross:

"More to the Aspartame Symposium: NutraSweet From the Wednesday Journal of December 18, 1991. Sue Ross, Manager, Public Relations The NutraSweet Company Deerfield, Illinois:

"Barbara Mullarkey's November 20 column reporting on the so-called aspartame symposium at the University of North Texas represented only one side of the aspartame story, as well as omitting some very key facts.

"First, there is an abundance of evidence showing that the organisers of the meeting never intended it to be a balanced, thoughtful inquiry into product safety. For example, The NutraSweet Company's chief medical spokesman, Dr. Robert H. Moser (Letters, December 11), was specifically asked not to attend. I was asked to speak, but only given twenty-four hours to accept the invitation because I was told the meeting program was on its way to the printer the next day.

"Additionally, I was later told by an official in the public affairs department at the university that the meeting organisers were told that they had to invite a NutraSweet spokesperson to balance the program. The company was given only fifty minutes to rebut more than eight-hours' worth of opposing presentation.

"Ms. Mullarkey omits another fact when she says I refused to answer questions. I stated in my talk that I would be happy to answer any and all questions that people had about our product – simply not in that meeting setting. I left a stack of my personal business cards and the business cards of the company's consumer centre, which has a toll free number. It was clear that the meeting organisers and speakers were not interested in our point of view or the facts that support aspartame as a safe food ingredient.

"I have enclosed a copy of my presentation with this letter. I would be pleased to send it to any Wednesday Journal reader who would like an accurate report of my presentation."

As I read such articles about the symposium and aspartame we'd held, I was struck anew by the veil of secrecy that has surrounded the diet sweetener. I could not help thinking that my own research showed that as early as 1973, the truth about aspartame had been sequestered from mainstream America. Through obscure labelling, aspartame had been concealed in cereal boxes, in laxatives, within children's chewable vitamins, alongside sugar and corn syrup in regular foods, in places you'd never think to look.

What worried me more than one company seeking to exert damage control was a question that constantly swirled around in my head: Who has the ultimate responsibility to protect the public? Thousands of people reported aspartame as harming them. There were four deaths because of aspartame. If I were a family member of someone who died from aspartame, I'd be furious! I almost lost my thyroid gland because of aspartame. I could have lost my life. That makes me even more furious! Whose responsibility was it to begin affirmative action?

Having lost faith in such government agencies as the FDA and the companies profiting by aspartame, I came up with only one answer: we the public must take action ourselves.

BITTERSWEET MONSTER

Although I had been researching aspartame for months, I learned a lot I did not yet know at the symposium. I also realised that I still had a long way to go before I would know everything I needed to about aspartame.

From gathering articles, reading reports, talking to clinicians and people who suffered its effects, I came to consider aspartame, simply, sweet poison. With my environmental background I knew that excessive exposure to poisonous chemicals inside the body can lead to toxic poisoning and weakened organ function. Armed with knowledge of the background of aspartame, I spent much of my free time at the university library trying to kind out about the chemical poisons that were in aspartame. I already knew that aspartame consisted of aspartic acid, phenylalanine and methanol, what I did not know was to what extent they effected one's health.

Aspartic acid and phenylalanine are two isolated amino acids. Amino acids make up proteins, and proteins allow humans to sustain life. There are approximately twenty-nine commonly known amino acids that account for hundreds of different proteins present in all living things. In order for a protein to be complete, it must contain all of its particular amino acids. The liver produces about 80 percent of the amino acids human beings need to maintain health. The remaining twenty percent must be obtained from outside sources. I learned that amino acids should not be taken in singular form, or isolated from one another. They are meant to be taken in combination, because, in nature, amino acids work together to build proteins in the body. Amino acids taken in isolation compete with one another, especially for entry into the brain. Chemicals not meant to enter the brain can

113

get in via an isolated amino acid. Food chemicals, such as artificial sweeteners like aspartame, piggyback on amino acids to gain passage into the control centres of the brain.

Books on nutrition and protein were piled high on my table by this point. Flipping from page to page, book to book, I read about the role amino acids play in the brain. The central nervous system cannot function without the proper balance of amino acids. Combined amino acids serve as neurotransmitters. They are necessary for the brain to receive and send messages. But, unless all the amino acids are present together, the proper transmission of the message can be inhibited. The amino acids in aspartame send a false signal of "sweetness" to the brain.

I discovered the two isolated amino acids in aspartame are fused together by its third component: deadly methanol.

I also learned that the methanol bonds the two amino acids together. When released, the methanol becomes a poisonous free radical.

Before I got too involved in learning about methanol, I wanted to continue reading up on the two amino acids so I would have a clear picture in my mind of what they were and how they functioned. I read that aspartic acid, an isolated amino acid, makes up forty percent of aspartame. Aspartic acid increases stamina and helps fight fatigue. But, under excess conditions, aspartic acid can cause endocrine (hormone) disorders. Hormones are regulated in the thyroid gland, which acts as the command centre for Graves' disease. It sounded to me like excessive aspartic acid must have played a big role in my illness. It was a scary thought.

I continued researching and found out that aspartic acid is a neuro-exciter, which means it affects the central nervous system. The central nervous system cannot function without amino acids. Unless all the amino acids are present together, accurate transmission of messages cannot be completed.

All the articles I read about aspartic acid labelled it as neuro-toxic, which means it is toxic to the brain in high concentrations. As in several diet drinks a day, I thought. These articles also claimed that hyperactivity is stimulated when too much aspartic acid is consumed which is unhealthy for children, especially if they are high energy, or hyperactive. According to the information, aspartic acid also poses a significant threat to the developing nervous systems of foetuses, infants, and young children. Dr. Diana Dow-Edwards proved this in her research, I remembered.

In order to enter the brain, aspartic acid must penetrate the blood brain barrier (BBB), the membrane that protects the brain, especially the developing brain, from toxins that cause death of brain cells.

I was intrigued by this and grabbed a bunch of different books and articles that explained or referred to the blood brain barrier. Along with scientific information, there were many historical and cultural facts. I learned that the blood brain barrier is referred to in some cultures as "The Third Eye," and In India, women traditionally place a jewel on their forehead over the BBB, paying respect to the body's sacred opening to the brain. This information was interesting and refreshing; However, I was a woman on a mission, so I turned back to the more technical and scientific texts.

I found out that it is at the BBB that nutrients enter the brain. If something toxic tries to pass through, the brain stops it by squeezing tightly to prevent its passage. However, if too many chemicals are ingested, no matter how hard the brain tries to prevent their entry, the overload will eventually pass through the BBB and into the brain.

Like other chemicals and nutrients, aspartame enters the brain at the BBB. Putting the pieces of information together like a puzzle, questions arose in my mind. I wondered how many headaches started because the BBB would try to prevent the two isolated amino acids in aspartame from carrying methanol into the brain. I thought back to the days of those blinding migraines I used to get after finishing a can of diet cola. I also happily realised I hadn't had one headache since I had stopped using aspartame!

All my research made me aware that the BBB was merely doing its job by trying to prevent an overload of phenylalanine, aspartic acid, formaldehyde and free methanol from bombarding my brain. Despite my efforts at keeping physically fit and healthy, I never really gave my body credit for doing what it was supposed to do: keep contamination from entering the human body's most integral organ – the brain.

I had heard of Dr. John Olney's work. Now I read it first hand. I saw the proof that his research confirmed, the frightening fact that even though it didn't belong there, aspartame penetrated the brain. Dr. Olney, research psychiatrist at the Department of Psychiatry at the Washington School of Medicine, began his research on the safety of aspartame in 1970. He informed G.D. Searle, the American developer of aspartame, that the aspartic acid in aspartame caused holes in the brains of laboratory mice. I winced at the image created in my head by Olney's report. However, I was bothered even more by the thought of this same toxic material floating around in the brains of adults – and even more children.

I read on. Dr. Olney determined that excess aspartic acid in the brain destroyed neurones. He brought up the fact that approximately seventy-five percent of the neural cells in any particular area of the brain can be killed before any clinical symptoms of a chronic illness are even noticed. By then it's too late to reverse any damage. "Neural cell damage can occur from excessive aspartic acid allowing too much calcium into the cells, triggering excessive amounts of free radicals, which kill brain cells," Dr. Olney wrote.

Other books on the subject revealed that excessive aspartic acid can be referred to as an "excitotoxin" because in excess it excites or stimulates neural cells to death. Many diseases are related to long-term exposure to excitatory amino acids.

I would later read a book by Professor of Neurosurgery at the Medical University of Mississippi, Dr. Russell L. Blaylock, MD. His book, 'Excitotoxins. The Taste that Kills,' would explain the damage caused by excessive use of aspartic acid. He would provide over 500 scientific references proving that excess free excitatory amino acids, aspartic acid being one of them, causes serious chronic neurological disorders along with a myriad of acute physical symptoms.

The risk to infants, children, pregnant women, the elderly and people with certain chronic health problems caused by excitotoxins is great. The Federation of American Societies for Experimental Biology (FASEB) stated, "The existence of evidence of potential responses... between males and females suggests a neuro-endocrine link and should be avoided by women of childbearing age and by individuals with affective disorders."

I discovered that aspartic acid from aspartame should be aided as a dietary supplement by pregnant and lactating women, and by infants and children because it has been shown to cause birth defects.

The exact route an adverse reaction will take in response to an excess of aspartic acid was under debate. However, as reported by the FDA, adverse reactions in their files included some of the following symptoms: headaches/migraines, nausea, abdominal pains, fatigue (because it blocks sufficient glucose entry into the brain), sleep problems, vision problems, anxiety attacks, depression, and asthma/chest tightness.

One of the most common complaints against aspartame is memory loss, I learned. Ironically, in 1987, aspartame manufacturer G.D. Searle began searching for a drug to combat memory loss caused by excitatory amino acid damage. On that note, I thought, maybe I should move on to component number two: phenylalanine.

L-phenylalanine is the other isolated amino acid found in aspartame. It constitutes fifty percent of the artificial sweetener. There are two types of amino acids, the D and L series, one being the mirror image of the other. Both the D and L forms of phenylalanine (L-phenylalanine and DL-phenylalanine) make up proteins. DL-phenylalanine should not be used by pregnant women or diabetics, in cases of high blood pressure, or anxiety attacks because it acts on the central nervous system.

I read that phenylalanine has been documented to cause seizures, elevate blood plasma and brain phenylalanine levels, alter the normal metabolism of pregnant women to three times the preconception level, cause severe retardation with IQ levels as low as twenty, and is responsible for a variety of birth defects. Excess phenylalanine blocks the production of serotonin, causing PMS symptoms, insomnia, mood swings and an increased craving for carbohydrates. It was depressing to learn that on top of these problems, phenylalanine's toxic effects are cumulative, so symptoms may not show up in short-term testing.

I studied the laboratory tests done and learned that phenylalanine's effects on rodents are different than on human beings. Phenylalanine raises rodent tyrosine levels, which serve as an antidote to phenylalanine's damaging effect in the rodent brain. Humans do not have this antidote generating ability. In essence, rodents don't respond to high doses of phenylalanine in the same way humans do. When humans consume any dose of phenylalanine, brain phenylalanine levels rise higher than the tyrosine levels.

Once again I felt the same horror I had felt the first time I'd heard Dr. Olney explain that rodents resist the effects of phenylalanine (fifty percent of aspartame) more efficiently than humans do, yet rodents still developed holes in their brains from aspartame. Now I saw the facts and figures. He had not exaggerated. He concluded, phenylalanine and aspartic acid have greater impact on humans than on rats. This was a frightening thing to learn, considering the awful effects aspartame was shown to have on laboratory rats.

One Saturday, I'd done all the chores: cleaned the house, washed so many loads of laundry I'd lost count, made an exhausting trip to the grocery store, and even washed the dog. The last thing I wanted to do was cook. So, I took the kids out for a hamburger at a fast food restaurant. As we stood in line behind a bus load of Boy Scouts shouting their orders all at once, I passively watched the people around me in the restaurant. It seemed like everyone was ordering diet colas. The kids looked over at me with a threatening gaze, "Don't say anything, Mom!"

"Shhh. Don't worry. I won't say anything," I responded to their fears of embarrassment. But I wanted to tell everyone in line, especially the children, "Don't order a diet cola. If only you knew what was in it!"

Consuming aspartame along with carbohydrates like a double burger, large fries, and a diet drink can lead to excess levels of phenylalanine in the brain. Phenylalanine blocks the production of serotonin in the brain, which causes an increased craving for carbohydrates, like another double burger, large fries, and diet drink. The whole thing becomes a vicious cycle.

When we returned home and I'd gotten the kids settled, I headed straight for my office. Where is that report I recently read about carbohydrates and aspartame, I wondered. It was like searching for a needle in a haystack, but I thumbed through mounds of paperwork until I finally found what I was looking for. "Yes," I voiced with enthusiasm. "I found it! Dr. Louis Elsas from Emory University wrote about this. Aspartame increases a craving for carbohydrates and sugar."

Hastily I scanned the papers. In his testimony before the United States Senate on November 3, 1987, Dr. Louis Elsas, M.D., illustrated how high levels of phenylalanine in the blood became concentrated in the brain. "This proves especially dangerous for foetuses and infants," he testified. He also demonstrated how phenylalanine was processed more efficiently in rodents than in humans. "Rats processed aspartic acid (forty percent of aspartame) and phenylalanine (fifty percent of aspartame) better than human beings did," Dr. Elsas wrote.

Excessive levels of phenylalanine in the brain have been shown to cause serotonin levels to decrease. This can lead to emotional disorders such as depression. Phenylalanine levels also increased significantly in people who continuously used aspartame. Even an isolated use of aspartame was shown to raise phenylalanine levels in the blood enough to be detectable. It made sense to me when I thought about how PMS, insomnia, mood swings and depression, among others, are all contemporary epidemics.

Excess phenylalanine levels in the brain block the production of serotonin, which cause PMS symptoms, insomnia, mood swings, and an increased craving for carbohydrates and sugar. So, I thought, more sugar cravings, more diet drinks, more depression, more prescribed medication. More carbohydrates, more weight gain, more depression, more diet products, more prescribed medication. A never-ending cycle that victimises the unsuspecting diet cola drinker.

Gathering all my papers I wandered onto the patio to finish reading. I stopped in the kitchen along the way to grab a cold bottle of water. A chivalrous old oak tree shielded the yard from the hot Texas sun like an enormous green umbrella. Its corps of red birds swooped from branch to branch while I saturated myself in research reports. The dog followed me outside and flopped lazily at my feet, unaware of the complicated materials I was studying.

Although I didn't have his book at the time, I would soon read Dr. Blaylock's work on excitotoxins. He would point out that early studies measuring phenylalanine levels in the brain were flawed. "Scientists who correctly measured 'specific' brain regions opposed to 'average' regions all noticed significant rises in phenylalanine levels," Blaylock would note in his book, 'Excitotoxins: The Taste that Kills'. "Specifically in the hypothalamus, medulla oblongata and the corpus striatum areas of the brain, which are the emotional control centres in the brain, these areas showed the largest increases in phenylalanine uptake."

The doctor would also write that excessive build-up of phenylalanine in specific brain regions could cause schizophrenia and increase susceptibility to seizures. In these cases, long-term use of aspartame could create a need for antidepressants, such as Prozac, to control schizophrenia, seizures, mood disorders and depression. As a matter of fact, Prozac use has increased to its highest levels ever. Health magazine's June 1990 issue wrote that "there has been a 102% increase in prescriptions for antidepressants over the past five years."

I had done a good amount of research by this time, but still I wasn't satisfied. I still needed to know about its third component: methanol. I was stunned to discover that methanol (wood alcohol or methyl alcohol) comprises ten percent of aspartame. Signs of methanol poisoning include lethargy, confusion, leg cramps, back pain, severe headaches, abdominal pain, impaired circulation, fainting, visual loss, laboured breathing and death.

As an environmental engineer, I am also certified as a hazardous waste and emergency response specialist and HAZMAT engineer. As such, it's my duty to stay abreast of human exposure to hazardous chemicals.

Sweet Poison

Knowing methanol was a hazardous chemical, I looked it up in my Emergency Response Guidebook and the National Institute for Occupational Safety and Health's Pocket Guide to Chemical Hazards.

I read that methyl (wood) alcohol is not only poisonous, it may be fatal if inhaled, swallowed, or absorbed through the skin. Physical contact with methanol can cause burns to the skin and eyes. Methanol is a flammable/combustible, colourless liquid with a characteristic pungent odour. It is a strong oxidiser, too. This requires strong antioxidants to counteract the toxic effects. Target organs for methanol poisoning are the eyes, skin, central nervous system and the gastrointestinal tract.

"Unbelievable," I muttered half under my breath as I slipped back into the house and placed the book on a shelf in my office to continue later. In the next few days, I read as much as I could about the methanol in aspartame from the various books and articles I had on loan from the library. The methanol in aspartame is released in the small intestine when it encounters the enzyme chymotrypsin. It is also "free form," so it immediately is absorbed into the bloodstream. As a matter of fact, the absorption of methanol is sped up considerably when it is ingested in "free form." Methanol found in natural foods is always in combination with ethanol or ethyl alcohol. The ethanol counteracts the dangerous effects of methanol as it is metabolised. In every case, natural ethanol is present in much higher amounts than methanol. Ethanol serves as an antidote for methanol toxicity in human beings. But, there is no ethanol in aspartame. No antidote present to stabilise the poisonous methanol.

Methanol's function in aspartame is to bond the phenylalanine and aspartic acid molecules together. When aspartame is heated above 86 degrees Fahrenheit, the methanol breaks "free." This can occur when aspartame products are improperly stored, transported in hot delivery trucks, or cooked. Yet, in 1993, the FDA approved aspartame as an ingredient in foods heated above 86 degrees Fahrenheit.

Above 86 degrees Fahrenheit, methanol breaks down into formic acid and formaldehyde, both listed in the National Institute of Occupational Safety and Health Pocket Guide as hazardous chemicals. Oh my God, I thought, formaldehyde is used in embalming fluid! Formaldehyde is used to embalm the dead!

"Formaldehyde is a dangerous neuro-toxin considered a cumulative poison due to its low rate of excretion once it is absorbed in the body," I continued to read. "Accumulating slowly and without detection, it is a known carcinogen, causing retinal damage, interfering with DNA replication, and causing birth defects. Its common symptoms include irritation of the eyes, throat and nose, burning in the nose, coughing, bronchial spasms and pulmonary irritation. Visual problems associated with acute toxicity do not always surface under chronic conditions, but appear as misty vision, tunnel vision, blurred vision, or obscured vision. Pancreas inflammation can occur with violent stomach pain, cramping and diarrhoea."

Putting down the document, I felt disgusted. Even with this information available to them, the FDA approved aspartame as an ingredient in foods! I wanted to stop reading, but couldn't, and picked up another document on the consumption of methanol. In it, I learned that the recommended limit for methanol consumption is 7.8 milligrams per day. One litre of an aspartame-sweetened beverage contains about fifty-six milligrams of methanol. Heavy users of aspartame consume as much as 250 milligrams per day of methanol, thirty-two times the Environmental Protection Agency's limit.

I turned to another article written by Dr. Woodrow Monte, director of the Food Science and Nutrition Laboratory at Arizona State University. He was so concerned about the unresolved safety issues regarding aspartame that he filed suit with the FDA in 1985 requesting a hearing to specifically address the methanol issue. He asked the FDA to "slow down on this soft drink issue long enough to answer some of the important questions." He wrote, "It's not fair that you are leaving the full burden of proof on the few of us concerned. And, for those who have such limited resources."

He continued by stating, "You must remember that you are the American public's last defence. Once you allow usage of aspartame, there is literally nothing I nor my colleagues can do to reverse the course. Aspartame will then join saccharin, the sulfiting agents, and God knows how many other questionable compounds enjoined to insult the human constitution with governmental approval."

I remembered that before this, in mid-1983, Commissioner Hayes approved the public use of aspartame in carbonated beverages and then resigned as FDA commissioner.

Dr. Monte's words rang in my mind: "You must remember that you are the American public's last defence." I thought back to the symposium held at school and the people who were involved.

During the national symposium, guest speaker Jim Turner had spoken of Ocala, Florida resident Shannon Roth's experience with NutraSweet.

Sweet Poison

Ms. Roth claimed NutraSweet and Equal caused irreversible blindness in her left eye. On Roth's behalf, Dr. Morgan Raiford of the Atlanta Eye Clinic wrote Senator Howard Metzenbaum during the Senate hearings of 1987 outlining the toxicity of aspartame in respect to Roth's blindness. He specified some of the serious side effects from using NutraSweet products which contain aspartame and recommended aspartame not become available to the public.

Dr. Raiford examined Shannon Roth in 1986. He observed evidence in her eye comparable to effects he observed in patients who suffered methyl alcohol (methanol) toxicity. Roth had damage in the central optic nerve fibres losing 225,000 of 137,000,000 optic nerve fibres. The extent of the damage caused partial blindness.

Dr. Raiford stated in his letter of diagnosis that the macular area of the eye's retina and optic nerve fibres were highly reactive to the toxicity of methyl alcohol because those fibres and nerve cells required from four to six times as much oxygen and nutrition as the visual pathway's peripheral nerves. These fibres and nerves, or rods and cones, could not go without oxygen for more than ninety seconds without some vision loss.

According to Dr. Raiford, "Aspartame is broken down by the upper portion of the digestive tract and methyl alcohol is then absorbed into the blood vessels of that region, which then travel through the body's entire vascular system. This toxic effect causes extensive damage to the human vessel pathway, ending with optic nerve atrophy as well as retinal starvation and visual loss. The intensity of the toxins from the methyl alcohol has a 'spin-off' effect, impairing nerve damage that varies with intensity to the amount of methyl alcohol the individual absorbs."

Ms. Roth used products containing aspartame for two years. During that time period, she experienced constant headaches, dizziness, insomnia, malaise, memory loss and depression. Every morning, Roth drank four to five cups of coffee with two packets of Equal in each cup. She drank diet cola throughout the day, and, in the summer, five to six glasses of an iced tea mix, one of whose ingredients was NutraSweet.

Roth confessed, "I innocently ingested these drinks thinking aspartame was the best thing since the wheel. Now I know it's a poison."

I remembered Jim Turner recounting to the symposium audience how Roth noticed a black spot in the centre of her eye while looking in a mirror. Four days later, she found herself in total blackness. She was admitted to the hospital where she underwent numerous X-rays, a spinal tap, and a CAT scan. The diagnosis: retrobulbar optic neuritis, or "blindness in one eye."

Dr. Raiford stated, "Every eye reacts differently to the amount of methyl alcohol it absorbs. Damage may not be noticed immediately, or as in the case of Shannon Roth, each eye can be damaged differently." Raiford stressed this important point: "In Shannon Roth's case, the damage in her eye was identical to damage I have repeatedly observed in the eyes of individuals whose eyes had been damaged by methyl alcohol toxicity."

Dr. Orion Ayer, Jr. also wrote the Senate subcommittee on Roth's behalf. Dr. Ayer wrote, "All her medical tests show no evidence of an associated neurologic deficit. I confirm that Roth had been using aspartame in relatively large doses during the time she experienced visual difficulties.

"In reviewing the literature available," Dr. Ayer continued, "the presence of methanol even in relatively small quantities can cause neurologic damage in susceptible individuals." Dr. Ayer also pointed out that the original investigators instrumental in the FDA approval of aspartame maintained NutraSweet was harmless. "But, certainly, most food additives including MSG and saccharin were thought to be harmless at the time of their introduction," he stated. "Ms. Roth's particular case may indeed be one of many patients who are sensitive to this additive."

I thought about Dr. Monte's words again. I thought about Shannon Roth, too. The FDA was her last defence and now she is blind. They failed her.

As depressing as it was, I had to keep reading and investigating because I knew that aspartame had to be linked to other types of damage besides eye problems.

I began uncovering interesting information on the association between diabetes and aspartame. However, I found that it was a relationship that lacked medical attention and investigation – not because it wasn't a problem, but primarily because of an under supply of information. I knew that people of any age can become diabetic. Although it is more common in older adults, diabetes is one of the most common diseases in children within the United States today. Approximately 127,000 children and teenagers nineteen years old and younger have developed diabetes.

I knew about diabetes because my adopted sister is an insulin-dependent diabetic. I had watched Beth slip into insulin-dependent diabetes like she was falling down a well.

Diabetes affects millions of Americans every year. It is a leading cause of death and disability, loss of eyesight, and kidney disease. The cost of treating complications of diabetes is very high and soaring. According to the American Diabetes Association, the direct and indirect costs of diabetes in the United States is over $92 billion per year.

Aspartame is allowed as a free exchange for most diabetics. H. J. Roberts, MD studies complications associated with aspartame in diabetics at the Palm Beach Institute for Medical Research in Florida. In response to a nation-wide survey, he encountered the following problems with diabetes related to aspartame: aspartame precipitated clinical diabetes; noticeably poorer diabetic control in insulin-dependent diabetics and diabetics using oral drugs to regulate insulin; more frequent hypoglycaemic reactions; and aggravation of diabetic complications, especially retinopathy, cataracts, neuropathy and gastroparesis.

Dr. Roberts believes that diabetic reactions are due to phenylalanine excess, altered neurotransmitters, the effects of methyl alcohol and an associated weight gain.

"The need to re-evaluate the safety of aspartame in diabetics is urgent," said Roberts, "especially during pregnancy and childhood. Its use by patients with symptomatic reactive hypoglycaemia and by close relatives of diabetics should be discouraged."

Diabetics do not need to use aspartame. My sister told me, "I'm an insulin-dependent diabetic, and I don't touch the stuff anymore! I have been healthier since I stopped using any NutraSweet. As a matter of fact, when I stopped using all aspartame, I decreased my insulin intake."

I wondered what else diabetics could use as a sugar substitute. I read more and discovered that there are other alternatives to chemical sweeteners. I learned about stevia, its uses, and even some history. Stevia is one of the most health restoring plants on the earth. Native to Paraguay, stevia has been used for centuries by South American Indians as a digestive aid and a topical salve for wounds. Stevia is a small green plant with leaves that are refreshing to the taste. In addition, stevia is thirty times sweeter than sugar. Stevia's leaves contain proteins, fibres, carbohydrates, iron, phosphorus, calcium, potassium, sodium, magnesium, zinc, rutin, vitamin A, vitamin C and an oil containing fifty-three other wholesome constituents. Because of its carbohydrate content, stevia reduces hunger and cravings for sweets and fatty foods. Hunger is reduced when ten to fifteen drops are used twenty minutes before meals. Research shows stevia may actually reset the hunger mechanisms where the pathway between the hypothalamus in the brain and the stomach is obstructed. This would help in feeling full sooner, and in eating less.

Stevia is non-toxic and has virtually no calories. It is one hundred per cent natural with nothing artificial or chemically added, is not "manufactured," is an anti-cavity and plaque retardant, and is rich in nutrients. Reports show stevia aids diabetics by lowering blood glucose levels, driving them closer to normal. I even read that some South American countries actually use stevia to aid diabetics! It also benefits hypoglycaemia by regulating blood sugar. Stevia has a "tonic-like" affect, increasing energy and mental stamina. Stevia has benefits for those affected by hyperactivity due to its nutritional qualities and health benefits. Studies show stevia lowers elevated blood pressure, yet does not affect normal blood pressure. Stevia inhibits the growth and reproduction of some bacteria and other infectious organisms, including bacteria that cause tooth decay and gum disease. Stevia is used as a mouthwash and can be added to toothpaste.

The FDA has not approved stevia as a food additive. It is "generally recognised as safe" for use as either a dietary supplement or an ingredient in a dietary supplement. This seemed odd to me since stevia had been used as a sweetener in diet sodas in Japan for years and years and had been widely accepted in South America for over 1,500 years.

Saccharin is another artificial sweetener that is an alternative to aspartame. After more than one hundred years of use, there have only been six complaints against saccharin registered with the FDA. Cases of human cancer sourced to saccharin appear to be rare and inconclusive.

I looked into saccharin's background and got a little history lesson on the product. Saccharin was first introduced in 1879! It is a crystalline compound unrelated to carbohydrates, with little nutritional value. Saccharin was originally extracted from the root of a plant growing in China. Saccharin is three hundred times sweeter than sucrose. When saccharin is heated to decomposition, it does emit some toxic fumes of nitrogen and sulphur oxides. However, these fumes are less harmful to the brain than formaldehyde from the methanol in aspartame or elevated blood phenylalanine levels.

In 1969, experiments were conducted in which laboratory rats were fed the equivalent of 800 cans of diet soda with saccharin per day for their entire lives. They developed bladder cancer. At this same time, John Olney's lab rats formed holes in their brains from a fairly normal dose of aspartame.

The saccharin experiments on rats have always been quite controversial. A panel of research scientists gathered at Duke University to review saccharin research. They reported: "Saccharin administered to rats at high dosages produced profound biochemical and physiological changes which did not occur in humans under normal patterns of use." The panel concluded that the appearance of tumours in the rats seemed to be a species and organ-specific phenomenon for which there was no explanation. The results of the study supported the fact that the present level of saccharin use as a food additive presents an insignificant cancer risk.

Saccharin was used during the sugar shortages of both World War I and II, especially in Europe. Several generations in the United States have used saccharin for decades. People with a low-calorie and low-carbohydrate diet, such as diabetics, had adopted saccharin as a part of their lifestyle for over a century.

Saccharin, like aspartame, has been one of the most researched food additives in history. It's been a research war. One article suggested the average user ingests less than one ounce of saccharin each year, yet five servings of aspartame for a fifty-pound child easily exceeds the RDA-recommended "allowable daily intake." This amount would raise the blood phenylalanine levels three hundred to four hundred percent. Evidence from years of research shows saccharin as safe. All human research has proven saccharin totally innocuous at human consumption levels. No research on human beings has shown an association between saccharin and cancer. Over thirty human studies have been performed, and they all resulted in saccharin safety. Some researchers believe a compound found in rats, not in human beings, may be the reason for the bladder tumours.

I read a scientific document that explained that saccharin is not metabolised when it is ingested and instead passes through the body unchanged – unlike aspartame, which metabolises into a long list of harmful toxins. I also read that saccharin does not react with DNA, the nucleic acid present in every living cell. This showed that saccharin lacked two of the major characteristics for a typical carcinogen.

In compliance with the Delaney Clause in 1977, the FDA tried to ban saccharin from the market. There was a public outcry over the ban and over the abuse related to the laboratory testing. At that time, Congress passed a moratorium preventing the FDA ban. The moratorium was extended six times because of the scientific research proving saccharin safe and consumer pressure to keep saccharin on the market. In 1985, FDA Commissioner Frank Young proposed at a Senate hearing to extend the moratorium again, stating that the FDA was less concerned about saccharin now than in 1977. He stated that the actual risk, if any, concerning saccharin appeared to be slight. It wasn't until 1991 when the FDA finally withdrew the proposed ban. They required any product containing saccharin, nonetheless, be labelled with a warning of a cancer risk. The moratorium will remain in effect through May 1, 2002.

I read that many soft drink companies have expressed interest in manufacturing more products with a blend of aspartame and saccharin. Many research scientists and medical professionals believe this would be a mistake. Dr. Jim Bowen's research demonstrates the hazards of combining these two artificial sweeteners. According to Dr. Bowen, aspartame and saccharin chemically combine to reach temperatures above boiling point. A blend of aspartame and saccharin could scald your bladder!

In the 1970s with the focus on the saccharin controversy, the FDA also banned cyclamate. The Washington Post on May 16, 1989 quoted the director of the FDA Toxicological Services Centre for Food Safety and Applied Nutrition,

Robert Scheuplein, as stating, "The decision to ban cyclamate was more a matter of politics than science. Meetings were not held. Things were not pursued. Work was not done. The people who were involved at the time were inadequate to the job."

Sorbitol is another alternate sweetener. Sorbitol is found naturally in fruits. It is used in many sugar-free gums, breath mints, and sugar-free candies. Too much sorbitol can cause digestive problems, however, resulting in bloating, cramping, or diarrhoea. A study of three-year-old children showed that a consumption of greater than 0.5 grams of sorbitol per kilogram of body weight caused diarrhoea.

Diabetics do have alternatives to using aspartame. They are just not as widely available as the products containing aspartame.

After all my research, I was sure that the components of aspartame could lead to a number of serious health problems, and not just for the diabetic. Symptoms can occur gradually, can be immediate, or can be acute reactions. According to Lendon Smith, MD, "There is an enormous population suffering from symptoms associated with using aspartame products, yet they have no idea why drugs, supplements, and herbs don't relieve their symptoms. Then, there are users who don't 'appear' to suffer immediate reactions at all. Even these individuals are susceptible to the long-term damage caused by excitatory amino acids, phenylalanine, methanol, and DKP."

Aspartame reactions are not allergies or sensitivities, but diseases and disease syndromes. Aspartame poisoning is commonly misdiagnosed because aspartame symptoms mock textbook "disease" symptoms, such as those for Graves' disease.

Aspartame changes the ratio of amino acids in the blood, blocking or lowering the levels of serotonin, tyrosine, dopamine, norepinephrine and adrenaline. It is typical for aspartame symptoms not to be detected in laboratory tests and on x-rays. "Textbook disorders" and diseases may actually be toxic loads resulting from chemical poisoning – aspartame poisoning." Mary Coleman, M.D. wrote in 1971. "Serotonin concentrations are markedly lower in the blood platelets of hyperactive children than in average kids. Hyperactivity decreases as serotonin levels increase." Aspartame causes a decrease in serotonin levels in the brain. If children do not eat a diet rich in vitamins and minerals and consume products containing aspartame every day, their serotonin levels will remain dangerously low, stimulating hyperactivity and aggravating depression. Giving an overactive child sugar-free aspartame products is fuel for the fire.

I now knew that even though short-term use of aspartame may not produce noticeable illness, the effects of aspartame poisoning are typically cumulative, and after sixteen years of public use by millions of people, frightening statistics have only begun surfacing. I had a feeling that, in the upcoming years, the truth about the patterns of aspartame poisoning would finally be known.

A REVOLVING DOOR

People often don't want to believe the truth about aspartame. The most common argument is, "If the FDA approved aspartame, then it must be harmless. The government wouldn't allow something on the market if it wasn't safe!"

There was a time in my life when I would have made that same argument, but because of my research and personal experience, I could no longer accept it. I knew it was comforting and "safe" to believe that the government would protect us, but I came to realise that the system doesn't always work.

In addition to the frightening symptoms and illnesses caused by aspartame, I was shocked to discover how bad the ingredients in aspartame are. Nothing, however, prepared me for what I would find when I investigated how aspartame and NutraSweet came to be approved as food additives.

As I looked into the general method that the FDA uses when approving a product, I learned that, basically, what the FDA does is review other scientists' studies – scientists usually paid by the company requesting the approval. If the FDA has any questions about a substance, it's assumed they will be addressed and promptly answered – by those corporations. Corporate scientists present a brief summary on why their product or food substance is safe for public use, hand the FDA corporate research results, fill out the appropriate paperwork, and wait for approval. Unfortunately, though this sounds good on the surface, reports can be biased, facts can be shaped, and objectivity blighted. The approval of cigarettes, silicon breast implants, thalidomide and the diet product "phen-fen" are some examples of the dangers of this system.

Sweet Poison

Once I had a little background on FDA practices, I began researching the story behind NutraSweet's approval. I found out that in 1973, when G.D. Searle Pharmaceuticals (mother company to the NutraSweet Company) first applied for a patent on NutraSweet as a food additive, they submitted their clinical tests to the Bureau of Foods. Although this may seem logical, aspartame was originally discovered as an ulcer drug. So, the Bureau of Foods recognised this and re-routed the research to the Bureau of Drugs for approval. Although G.D. Searle labelled NutraSweet a "food," its chemical components didn't change. The FDA luckily caught this and didn't let it go, no matter how the paperwork was filed.

Sitting back for a moment, I tried to absorb this information. I knew that if aspartame was resubmitted today for approval through the Bureau of Drugs, it wouldn't pass as a food additive. And yet, men, women, pregnant mothers, children and babies are consuming foods, medicines, vitamins and soft drinks containing a drug originally intended to treat ulcers. A drug which had its US patent revoked in 1974 due to the findings that it caused holes in the brains of lab animals. To me, this was unconscionable. So then, how did aspartame succeed in getting the 1981 patent approved by the FDA? I wondered. Placing the 1973 patent into the hands of the Bureau of Drugs suddenly changed the rules on G.D. Searle.

I learned that, routinely, the Bureau of Foods does not perform any testing, but the Bureau of Drugs does. According to a November 3, 1987 report from a Health and Safety Concerns Hearing before the United States Senate Committee on Labor and Human Resources, when the Bureau of Drugs reviewed the first clinical tests on aspartame, they found them unacceptable. The tests were returned to the Bureau of Foods with an explanation on how to conduct proper clinical testing. But, the Bureau of Foods doesn't conduct any tests. So, where did the NutraSweet Company, which wanted to use aspartame as an artificial sweetener, go from there?

Martha M. Freeman, MD, FDA Division of Metabolic and Endocrine Drug Products, wrote in August, 1973: "The administration of aspartame at high dosage levels for a prolonged period of time constitutes clinical investigation as a new drug substance."

The information submitted for the review was inadequate to permit a scientific evaluation of clinical safety. Dr. Freeman recommended an IND (notice of claimed investigational exemption for a new drug) be filed, including all required manufacturing controls and pharmacology and clinical information. She also recommended that marketing this chemical as a sweetening agent should be contingent upon satisfactory demonstrations of clinical safety of the compound.

On January 11, 1974, Richard Crout, MD, acting director of the FDA Bureau of Drugs to the Associate Commissioner for Compliance stated that the Bureau of Foods, Division of Toxicology had recently requested a Bureau of Drugs medical review of clinical safety studies for aspartame. The information on aspartame submitted for review at the time was limited to clinical summaries and to tabulated laboratory mean values. No protocols, manufacturing controls information, nor pre-clinical data had been provided by G.D. Searle to either bureau. Because there were substantial deficiencies submitted in each area of required information as a food additive, Dr. Crout recommended approval be conducted under the investigation of New Drug Regulations.

Despite these roadblocks, the FDA reversed itself, and aspartame was approved in 1981.

How? I asked myself.

Jim Turner, attorney for the Community Nutrition Institute, Washington, DC, who we had as a speaker at the national symposium, said in reference to the FDA's 1981 approval of aspartame, "When studies are done where the majority of the research is paid for by the industry, it is not scientists' integrity that is the problem. It is not the intention of anyone involved in the process that is the problem. It is the system itself that is the problem. The original issue becomes skewed in a direction where certain questions that would normally be asked by critics are not asked. In fact, most questions raised by critics are never funded for research. It is very important that both sides of an additive, and that the critics of an additive, have equal resources to do their research."

Turner continued, "This is the only way we can ever sort out whether a product like NutraSweet is effective or not; is safe or not! A substantial proportion of the studies that NutraSweet approval relied upon did not appear in peer-reviewed journals and were not peer-reviewed. This was one of the most important weaknesses in the record."

Aspartame succeeded in permeating the food supply when there was, and still is, an abundance of doubt from the scientific community. Although I had learned much history, the whole story was so muddled and confusing that I now began creating a time line of events leading up to aspartame's approval as I pored over books and articles. I wanted to be sure that I didn't miss a thing.

The flask that boiled over in 1965 during the testing of an ulcer drug by G.D. Searle chemist, James Schlatter, leading to the accidental discovery that aspartame had a sweet flavour, was, for the company, a happy accident, but it was only the first step.

In 1969, physician and biochemist Dr. Harry Waisman, professor of paediatrics at the University of Wisconsin and director of the university's Joseph P. Kennedy, Jr. Memorial Laboratories for Mental Retardation Research, approached G.D. Searle to conduct a study on the effects of aspartame on phenylketonurics (PKU). An expert on PKU, Dr. Waisman proposed to study the genetic disorder in response to aspartame – using primates.

"Of the seven infant monkeys fed aspartame mixed with milk, one died after three hundred days, and five other monkeys suffered grand mal seizures," Dr. Waisman reported. He submitted his findings to G.D. Searle, however, the test results were not passed on to the FDA. In 1975, an FDA investigative task force discovered Dr. Waisman's study and questioned Searle as to "why the study was deleted from FDA records." Searle officials could only respond that they didn't know why.

Before he could complete his work, Dr. Waisman was killed in an automobile accident in March of 1971. At that point, the FDA considered his research important, but as of 1980, dismissed his findings as "incomplete."

G.D. Searle then granted researcher Ann Reynolds funds to study Waisman's findings. According to congressional records, "Her findings were fragmented, as she evaluated plasma aspartic acid levels only; not the neuro-toxicity or seizure potentiality as did Dr. Waisman."

Primates were never again used in aspartame research. All successive studies submitted by Searle to the FDA were performed on rodents. Unfortunately, the tests on rodents were not as accurate as those on primates. Rodents had to be fed at least sixty times more aspartame to duplicate the intake effects on humans, which made the tests incomparable.

It was discovered at that time that the laboratory rats used in saccharin testing had been fed the equivalent of 800 cans of diet soda sweetened with saccharin per day. There was a public outcry over a proposed ban on saccharin and on the potential abuse related to the excessive amount of saccharin fed to the test animals. By 1970, the safety of saccharin was seriously being questioned. The FDA had banned cyclamate.

Independently, Dr. John Olney, a research psychiatrist in the Department of Psychiatry at the Washington School of Medicine, began his research on the safety of aspartame in 1970. The following year, Dr. Olney informed G.D. Searle that the aspartic acid in aspartame caused holes to form in the brains of his test mice. Ann Reynolds, the same researcher hired by Searle to investigate Dr. Waisman's findings, was again contracted by Searle to investigate Dr. Olney's findings. She confirmed his findings in a similar study.

Despite the problems with aspartame that laboratory testing had revealed, G.D. Searle petitioned the FDA in 1973 for the use of aspartame in all foods.

FDA Commissioner Alexander Schmidt, M.D. approved aspartame as a food additive in dry foods only in 1974. Prior to Schmidt's approval, Dr. Olney and Jim Turner met with G.D. Searle representatives to discuss Dr. Olney's findings. Searle representatives told Turner and Olney that Olney's studies "raised no health problems." Not surprisingly, Olney and Turner disagreed.

Searle did not notify the FDA of Dr. Olney's findings until after Dr. Schmidt granted approval for aspartame in dry foods. No tests similar to Olney's were ever submitted to the FDA by Searle.

In 1975, an FDA task force was formed to investigate aspartame safety concerns. The task force was headed by FDA lead investigator Philip Brodsky and FDA toxicologist Adrian Gross, M.D. Both men were charged with examining the original test material submitted by G.D. Searle. The task force's investigation included examining the data Searle had previously submitted to the FDA for the approval of their Copper-7 IUD. Searle eventually removed the IUD from the common market and lost a $9,000,000 lawsuit in 1988.

After several months of investigations, the task force submitted a 15,000-page document with a summary of over eighty pages. The summary states, "The task force uncovered serious deficiencies in Searle's integrity in conducting high quality animal research and in Searle's ability to accurately determine the toxic potentiality of NutraSweet. We found instances of 'irrelevant' animal research where experiments had been 'poorly conceived and executed, inaccurately analysed and reported.'" They also stressed, "The problems we found throughout the studies revealed a pattern of conduct that 'compromised the scientific integrity of the studies.'"

Sweet Poison

One of the most stunning findings in the task force report was that many of the test animals fed aspartame had developed large tumours. The tumours were cut out and the animals returned to the study. In some cases, the tumours were not examined for malignancy nor reported to the FDA. When the FDA questioned Searle about these irregularities, company representatives replied, "The masses were in the head and neck areas, preventing the animals from feeding." In several cases it was discovered by FDA investigators that some animals had actually died, but were reported by Searle researchers as alive in the files they submitted to the FDA.

The FDA concluded that some of Searle's studies were questionable. Commissioner Schmidt's earlier aspartame approval was suspended.

Dr. Gross said, "The sixteen-month investigation was at best irrelevant because the task force was limited to only analysing whether Searle had falsified the test data. We had been directed not to examine the method of testing since the FDA had already accepted the test procedures."

As a result of the task force's investigation, the first Senate Subcommittee on Labor and Public Welfare hearing was called on April 8,1976 to discuss aspartame and several other Searle drugs. Dr. Gross, Senator Ted Kennedy, FDA Commissioner Schmidt, and a task force appointed by Commissioner Schmidt were present at the hearing. Senator Kennedy, chairman of the Senate subcommittee, stated at that time, "The extensive nature of the almost unbelievable range of abuses discovered by the FDA on several major Searle products is profoundly disturbing."

Commissioner Schmidt said the FDA had never conducted such an examination as they did with Searle. He added, "The FDA investigation clearly demonstrates that in the case of G.D. Searle Company, we have no basis to rely on the integrity of the data submitted."

Schmidt continued, "We have teams within the FDA looking at Searle's drug research... Obviously, when you're out in the public saying you're going to approve aspartame and people come along with objections and say it's going to destroy people's brains or whatever, you take a pretty hard look at it.

"Immediately, there were objections raised," he continued. "Objections I hadn't heard before. Jim Turner and John Olney came out of the woodwork... By and large, the results of the Searle investigation were very disturbing. What we saw was a lot of shoddy work, and we saw studies that were far less than perfect. We stayed approval, pending a look at what people would bring the board of inquiry."

A five-member FDA task force headed by FDA Inspector Jerome Bresler was then given fifteen Searle studies to examine. Their conclusion: "Some of the test animals developed uterine tumours. When asked about the tumours, Searle admitted the tumours were due to diketopiperazine (DKP), a by-product of aspartame hydrolysis. The higher the dosage, the higher the incidence of uterine tumours."

The task force also determined that some of the blood tests had been tampered with. Searle claimed they experienced problems with their instruments, and, therefore, substituted the results of some studies with other test results. A Rockville, Maryland firm, Universities Associated for Research and Education in Pathology (UAREP), examined twelve of the fifteen studies. They reported finding written accounts of brain tumours.

In January of 1977, FDA Chief Counsel Richard Merrill formally requested that United States Attorney Samuel Skinner conduct a grand jury investigation on the tests submitted to the FDA by G.D. Searle. The investigation was based on allegations that G.D. Searle was "concealing material facts and making false statements in animal study reports concerning the safety of the drug aldactone and the food additive aspartame."

At that time, the FDA cited two specific NutraSweet studies needing special attention. The first was a primate study in which monkeys suffered seizures, but were never given autopsies. The second was a toxicity study on hamsters.

United States Attorney Skinner resigned from the case merely two months after commencing his initial aspartame investigation. He accepted a position with the law firm Sidley and Austin – the partnership defending G.D. Searle in these same investigations Skinner later returned to the federal government, accepting the cabinet position of Secretary of Transportation. His responsibilities included the transportation and handling of hazardous materials, including the import of aspartame shipped to the United States from G.D. Searle's Japanese supplier.

William Conlon, senior assistant United States Attorney, was appointed United States Attorney to replace Skinner. Conlon also took no action on the case and, like Skinner, also accepted a position with Sidley and Austin, Searle's law firm. Thomas Sullivan was then appointed United States Attorney, but he took no action in the investigation.

Sweet Poison

The statute of limitations for prosecution was set to expire in December of 1977. Indeed, the statute of limitations for a grand jury investigation had already expired. Time to take legal action against Searle was rapidly running out.

During this time, Donald Rumsfeld, former congressman and chief of staff for the Ford administration, was hired as G.D. Searle's president. Rumsfeld earned $2 million in salary and $1.5 million in bonuses between 1979 and 1984. As president of G.D. Searle Company, Rumsfeld immediately hired three government officials: John Robson, Robert Shapiro, and William Greener, Jr.

John Robson was hired as Rumsfeld's executive vice president. He was a former lawyer with Sidley and Austin, Searle's law firm, and also served as chairman of the Civil Aeronautics Board, working under the Department of Transportation.

Robert Shapiro was hired by Rumsfeld as general counsel. He went on to become the first head of Searle's NutraSweet Division. Shapiro was Robson's special assistant at the Department of Transportation.

William Greener, Jr. was hired as Searle's chief spokesperson. He was a former spokesman for the Ford administration.

In 1978, North-eastern Illinois University's Department of Psychology submitted a study on aspartame. They documented the following aspartame research results: reproductive dysfunction in both male and female test animals; endocrine dysfunction including the pituitary gland, the thyroid gland, the ovaries, and the testes; an increase in weight as a result of using aspartame; and a decrease in locomotor function.

In that same year, proceedings were held at the National Academy of Sciences of the United States of America. Research was presented showing elevated blood phenylalanine levels unfavourably affecting foetuses of mothers who carried the PKU gene. This caused a lower IQ and a higher incidence of developmental abnormalities.

In 1979, responding to an official request, United States Attorney Tom Sullivan wrote a formal statement to the FDA stating his reasons for not prosecuting Searle on aldactone. However, he made no mention of aspartame.

Several independent studies on aspartame were performed during 1979. As reported in Science magazine, studies linked methanol (ten per cent of aspartame) to foetal alcohol syndrome and diminished capacity in new-born rats. The New England Journal of Medicine published a study showing a high incidence of birth defects as a result of elevated phenylalanine levels in PKU women. Phenylalanine is fifty percent of aspartame. Dr. Daniel Azarnoff, head of G.D. Searle's Research and Development Division, made the statement, "Rats eating the required amount of DKP (diketopiperazine from aspartame) had a statistically significant number of tumours in their wombs."

At that time, before approving aspartame as a drug or as a food additive, the FDA requested a review of the substantial objections they were receiving concerning aspartame safety. Routinely, public hearings are held by administrative law judges, but in this case, the FDA suggested a public board of inquiry be assembled with three scientists rather than individual judges.

The public board of inquiry was empanelled in 1980. The board members were: Peter J. Lampert, MD, professor and chairman of the Department of Pathology, University of California, San Diego; Vernon R. Young, Ph.D., professor of nutritional biochemistry, MIT; and Walle Nauta, MD, Ph.D., institute professor, Department of Psychology and Brain Science, MIT. Dr. Nauta chaired the board.

The board's investigation had a touch of the bizarre. They were obligated to rely on the findings of the Rockville, Maryland, firm, UAREP (investigators from the 1977 task force) because FDA officials denied them access to the complete task force reports. After reviewing the file, the board voted unanimously to recommend banning aspartame for human consumption.

Dr. Nauta expressed the board's concerns about aspartame safety, "based on the fact taint the investigation did not include the 'validity' of Searle's tests, only their 'results.'"

The board concluded, "The available data on laboratory rats did not rule out the possibility of aspartame causing brain tumours, and that, indeed, the evidence suggested that aspartame might induce brain tumours."

After these recommendations were submitted, another team was assembled to study the public board of inquiry's finding. A five-member commissioner's team of scientists was empanelled. By then, so much data was being assembled for review, team members broke into groups and were given different review assignments.

Three team members investigated the brain tumour issue. The other two members were assigned brain damage, mental retardation, and endocrine problems. The three scientists reviewing the brain tumour studies expressed serious concerns with the data. They indicated aspartame as a causative factor in tumours. The other two scientists were satisfied with tests in the area they studied, indicating aspartame did not cause brain damage.

For reasons never announced, a sixth member was appointed to the team. With the addition of the sixth member, the vote became deadlocked. Three voted for the approval of aspartame, three voted against.

Jacqueline Verrett, Ph.D., toxicologist and senior member of the review team, was extremely critical of how the review was performed. "It was pretty obvious that somewhere along the line, the bureau officials were working up to a whitewash," she stated. "The Bureau of Foods under Howard Roberts either discarded or completely ignored the problems and deficiencies outlined by the team in their report. It is unthinkable that any reputable toxicologist giving a completely objective evaluation of the data resulting from such a study could conclude anything other than the study was impossible to interpret and worthless, and should be repeated as the safety questions still remain unanswered," Verrett concluded.

Dr. Verrett would testify before the third Senate hearing concerning aspartame safety in 1987.

The scientific journal 'Neurobehaviorial Toxicology' took up the question of aspartame safety in the article, "Brain Damage in Mice from Voluntary Ingestion of Glutamate and Aspartate." Their researchers stated, "The harmful effects of aspartate salt from aspartic acid, forty percent of aspartame, could not be detected after years of exposure except in the form of obesity or neuroendocrine disturbances, as are known to occur in rodents following treatment in infancy."

In September, G.D. Searle re-petitioned for aspartame approval. The FDA public board of inquiry denied approval pending further brain tumour testing. The board also formally revoked aspartame's 1974 approval granted by Commissioner Schmidt.

In January 1981, G.D. Searle again reapplied for aspartame approval. Arthur Hull Hayes, Jr., MD was appointed FDA Commissioner by President Ronald Reagan in April. In July, Dr. Hayes overruled the public board of inquiry's recommendation that aspartame "not be approved until further animal testing be conducted to resolve the brain tumour issue." Two FDA officials went on record, stating that they disagreed with Hayes' judgement on aspartame.

Hayes granted the approval for aspartame to be used in dry foods as NutraSweet, and as the table-top sugar substitute, Equal. Nonetheless, the law stated that "tests found inconclusive pertaining to a food additive cannot receive FDA approval." Apparently, aspartame was the exception to that rule.

Prior to 1976, more than ninety per cent of aspartame testing submitted to the FDA was performed by independent researchers, not by G.D. Searle researchers. Many of these scientists had been commissioned by the FDA. Hayes' predecessor, FDA Commissioner Goyan, had denied aspartame approval despite six years of Searle's failed attempts.

Even though it was a drug, the FDA approval of aspartame as a food additive in 1981 made G.D. Searle's NutraSweet exempt from continued safety monitoring. Searle was no longer obligated to defend against adverse reactions associated with NutraSweet nor required to submit safety reports to the FDA of such reactions. NutraSweet was on its way to becoming the world's favourite artificial sweetener.

What was spurring it on? The anticipation of its manufacturers. Big profits to be gained from the millions of people ignorant of any health threat and desirous of consuming sweetened foods and drinks devoid of calories.

BEHIND CLOSED DOORS

Much time had passed since I admitted myself to the hospital emergency room with a racing heartbeat and was diagnosed with Graves' Disease. While my marriage never returned to the way it was prior to my illness, I continued to focus on raising my children, teaching and speaking out publicly about the dangers of aspartame, and researching its effects. Despite all the hours I devoted to making myself an expert on the subject, there was always more to read and learn.

I added to my time line as I learned more of aspartame's history and current status. Another piece of startling information I found and integrated was that between 1979 and 1982, four more FDA officials associated with the approval of aspartame accepted jobs with NutraSweet companies: S.M. Page, associate chief counsel for Foods, Health and Human Services and special assistant to the FDA commissioner; Sherwin Gardner, FDA Deputy Commissioner; Mike Taylor, attorney with the board of inquiry; and Albert Kolbye, associate director for the Bureau of Foods for Toxicology.

In 1982, Senator Howell Heflin, chairman of the Senate Ethics Committee, proposed an amendment to change the laws in order to extend the United States patent for aspartame. Senator Orrin Hatch, chairman of the Labor and Human Resources Committee, led a discussion on the Senate floor supporting the proposed extension. Senator Byrd initiated an amendment to extend Searle's patent on Senator Heflin's behalf. Representative Henry Waxman sponsored the Orphan Drug Act. The ruling to extend NutraSweet's patent until 1992 was approved as an amendment to the Orphan Drug Act.

G.D. Searle presented campaign contributions to the following senators and representatives: Senator Robert Byrd, Democrat, West Virginia; Senator Orrin Hatch, Republican, Utah; Senator Howell Heflin, Democrat, Alabama; and Representative Henry Waxman, Democrat, California.

NutraSweet was approved for use in carbonated beverages and carbonated beverage syrup bases in 1983. FDA Commissioner Hayes resigned his position to accept a position as senior medical advisor to Searle's public relations firm, Burson Marsteller. Hayes' consultant fee was reported as $1,000 a day.

Anthony Brunetti, FDA Consumer Product Officer, drafted the proclamation to approve using NutraSweet in soft drinks. Brunetti later resigned and accepted a position with the Soft Drink Association as science advisor.

That same year, the FDA raised the acceptable maximum daily intake of aspartame from 20 mg/kg/day (milligrams per kilograms of body weight per day) to 50 mg/kg/ day, the equivalent of going from two cans of diet soda per day to seventeen cans per day for a 150 pound person. For a twenty-five to thirty pound child, 50 mg/kg/day could be reached merely by drinking three cans of diet soda or chewing a pack of sugar-free gum, plus taking a multivitamin with aspartame, eating cereal sweetened with aspartame, and a serving of sugar-free pudding, or ice cream containing aspartame. As early as 1976, the Journal of Toxicology and Environmental Health had reported children were consuming as much as 75 mg/kg/day. A two-litre bottle of diet soda contains 1,200 milligrams of aspartame.

In 1984, aspartame was approved for use in children's chewables and multivitamins. Woodrow Monte, Ph.D., director of the Science and Nutrition Laboratory at Arizona State University, stated, "There are presently no animal nor mammalian studies that have evaluated the possible mutagenic or carcinogenic effects of a chronic intake of methanol (ten percent of aspartame)." Dr. Monte requested an Arizona State hearing on a petition to ban products containing NutraSweet. He was not granted the hearing until April of the following year. He maximised his efforts to get Arizona to recognise the dangers of human consumption of aspartame, but did not succeed. The Arizona legislature had changed the laws without public notice barring state regulation of FDA approved food additives. The measure passed under a "toxic waste bill."

Between 1979 and 1984, G.D. Searle hired over a dozen lobbyists. United Press International traced nearly $200 thousand in campaign contributions including contributions to House Majority Leader Burton Barr and Representatives Don Aldridge, Karen Milt, and Jan Brewer. Searle hired the following lobbyists to help fight Dr. Monte's efforts to ban aspartame in Arizona: Andrew Herwitz, Arizona Governor Babbitt's former chief of staff; Charles Pine, Arizona lobbyist; Roger Thies, a Searle lawyer; and David West, a Searle official.

Richard Wurtman, MD at the Department of Brain and Cognitive Science at MIT, began research on NutraSweet.

Ronald Gautieri Ph.D. and Michael Mahalik, Ph.D. at the Department of Pharmacology at Temple University in Philadelphia demonstrated in their study that aspartame produced brain dysfunction new-born mice.

William Pardridge., M.D. from the University of California at Los Angeles, testified along with Dr. Richard Wurtman before a Senate Committee on Labor and Human Resources. The doctors urged that labelling requirements for aspartame be amended to include a limit on the quantity consumed. They brought up two important points: "Children may be at risk of suffering brain damage from excessive intake of aspartame, and aspartame consumed at the same time as carbohydrates, in carbonated soft drinks, would double the effect on the brain as aspartame alone."

In 1986, Dr. William Pardridge released a study in the American Medical Association of Scientific Affairs Review stating the two amino acids found aspartame, aspartic acid and phenylalanine, were neuro-toxic. He recorded a drop in IQ in babies born to mothers with elevated phenylalanine levels and noted a decrease in choice reaction time in adults with slightly elevated phenylalanine levels.

The quarterly report submitted by the Department of Health and Human Services showed three thousand complaints against aspartame reported to the FDA and the Centre for Disease Control (CDC). Most of the complaints related to brain function and to behaviour changes.

Food Chemical News reported that the Community Nutrition Institute had petitioned the FDA to ban aspartame, citing cases of seizures and visual problems.

Other countries around the world were issuing their own warnings. The Department of Psychology at Leeds University in Leeds, England, reported in the May 10 issue of the British medical journal, Lancet, that consumption of aspartame caused an increase in appetite and weight gain. Leeds researchers Blundell and Hill stated in their findings, "Aspartame increased motivation to eat and decreased the feeling of fullness in the test volunteers.

"Aspartame in some circumstances had appetite stimulating properties in comparison to drinking water," they cited. "After drinking aspartame, the test volunteers were left with a 'residual hunger' compared to similar feelings after eating glucose."

According to Blundell and Hill, "This residual hunger led to an increase in food consumption." They concluded that people using aspartame may receive ambiguous signals important for the control of appetite and ingestion. Blundell and Hill explained, "This confusion leads to a loss of control over appetite, particularly in vulnerable individuals of normal weight who are dieting and who may be consuming large amounts of dietary aids for weight control. This may contribute to disordered patterns of eating prevalent among certain groups of normal weight individuals."

In 1987, the NutraSweet Company petitioned for FDA approval to include aspartame as an ingredient in baked goods and baking products.

The General Accounting Office reported monitoring the food additive process for aspartame, and a third United States Senate hearing was summoned on "NutraSweet: Health and Safety Concerns." Senator Howard Metzenbaum chaired the Committee on Labor and Human Resources which oversaw the hearings. Many research scientists and medical researchers responded to the Senate hearings. (Included in the appendices are copies of the letters written by medical professionals to Senators Edward Kennedy, Howard Metzenbaum, and Orrin Hatch expressing their concerns about the dangers aspartame impressed upon human beings, especially pregnant women and foetuses.)

The Labor and Human Resources Committee heard testimony concerning the health and safety of NutraSweet on November 3. This hearing was important not only because it was the third time the United States government attempted to settle the aspartame issue once and for all, but it was the last hearing ever to be held questioning the safety of the artificial sweetener.

I saw, once again, the shocking story Mary Stoddard had told at the symposium about Air Force-Major Michael Collings testimony on the tremors and seizures he experienced from 1983 until 1985 as a result of drinking three quarts of lemon-flavoured powdered drink mix with

NutraSweet plus two or more diet sodas a day. It was just as chilling when I saw again how Collings closed his testimony by saying, "My career as a pilot is over. My concern is for others who consume NutraSweet, especially people who fly fighter-type aircraft and those in high risk jobs."

A new case I hadn't known of was the one concerning Larry Taylor, a nurse anaesthetist and a Senate panel member. He testified about his encounter with aspartame. In 1985, Taylor experienced one of three grand mal seizures minutes before giving a patient a spinal injection. "My neurologist suggested I stop my daily intake of four to six diet colas a day, plus numerous packets of artificial sweetener on my dry cereal every morning," he said. Taylor never experienced another seizure. His memory loss, eye problems, headaches and hives permanently disappeared.

Taylor concluded his testimony by stating, "My concern is for health professionals who ingest products containing aspartame and for their patients whose lives are in the professionals' care."

Jacqueline Verrett, Ph.D., former FDA toxicologist, spoke at the hearing, insisting that the original aspartame studies were "built on a foundation of sand." She stated, "The repeated flawed tests were the basis of FDA approval and were a disaster that should have been thrown out from day one." She submitted written testimony documenting her beliefs that aspartame was introduced into the market without even the most basic toxicity information.

According to Dr. Verrett, "There is no data to assess the toxicity of the interactions of diketopiperazine with other aspartame metabolites, other additives, drugs, or other chemicals."

Dr. Verrett had served as a member of the 1977 FDA Bureau of Food team summoned to examine the authenticity of three Searle aspartame animal studies. According to her written testimony, she stated, "I wish to emphasise... that we were specifically instructed not to be concerned with, or comment upon, the overall validity of the study; this was to be done in subsequent review carried out at the bureau level. It was apparent that the review, on a point-by-point basis, discarded or ignored the problems and deficiencies outlined in this team report, and concluded that, even in total, these problems were insufficient to render the study invalid. It also appears that the serious departures from acceptable toxicological protocols that were noted in the re-evaluation of these studies were also discounted."

The deficiencies and improper procedures Dr. Verrett referred to were:

• In the DKP study, the animal feed was unevenly mixed. The laboratory rats were allowed to choose between the standard animal feed and the larger particles of DKP. DKP is one of aspartame's major breakdown products.

• Animals that died were not reported and were instead considered "alive" for the records.

• Tumours were removed and undeclared. The animals were returned to the study.

• Many tissues decomposed before post-mortem examinations were performed.

• There was sporadic monitoring and/or inadequate reporting of feed consumption by the animals and of animal weights.

• No formal protocol had been written until the study was already underway.

Dr. Verrett stated, "Any one of these practices would negate a study of this kind. Any reputable toxicologist, in my opinion, would conclude the study was impossible to interpret and worthless. The safety of aspartame and its breakdown products have not been satisfactorily determined."

Alfred Miller, MD presented written testimony at the hearing. He opined, "As a physician, I have no NutraSweet in my home, and I encourage all of my patients with any tendency towards mood disorders and/or headaches to refrain from NutraSweet products. I would also assume that the toxic effects are only the tip of the iceberg. Should these patients be followed for a longer period of time, I feel some of the vascular type of headaches may progress to actual strokes with fatal outcomes."

Dr. Miller first became aware of the dangerous effects of NutraSweet when a long-time patient developed severe headaches for the first time. "She could not find a cause for the debilitating headaches nor determine a diagnosis," Dr. Miller confirmed. "This patient did not respond to over-the-counter medications she had been taking for her daily headaches and recent insomnia. After reading an article in the Journal of the American Medical Association about NutraSweet possibly causing headaches, I suggested this patient eliminate all NutraSweet and refrain from using headache medicines. Within twenty-four hours, the patient was headache-free and has remained so."

Dr. Miller continued, "I had similar successes with my other patients whose headaches stopped when they discontinued all use of

NutraSweet. I also had patients who experienced problems with mood swings, agitated behaviour, and short tempers whose behaviours returned to normal after stopping all NutraSweet."

Three scientific researchers testified before the Senate subcommittee: Dr. Louis Elsas, director of Emory University's Medical Genetics Centre; William Pardridge, MD, chief of staff at the University of California, Los Angeles Blood Brain Barrier Laboratory; and Peter Dews, MD, Ch.B., Ph.D., a Harvard Medical School professor of psychiatry and psychobiology. *

Dr. Elsas stated, "Aspartame is a well-known neuro-toxin. Aspartame ingestion in some as yet unidentified doses, will produce adverse effects both reversibly in the adult brain and irreversibly in the developing child or foetal brain. For the past twenty-five years, I have studied the prevention of mental retardation and birth defects related to the consumption of phenylalanine. I wanted to know why the following questions had not been answered:

• "Does an excess of phenylalanine occur with aspartame ingestion?
• "Will excess phenylalanine adversely affect human brain function?"

Dr. Elsas' impassioned statement that a two-to-five-fold increase in phenylalanine disturbs the body's balance of amino acids can result in serious health problems was followed by his strong words against NutraSweet, "It gives a false impression that NutraSweet is good for you, that it's nature's best, and that it might be good for children to take. I'm angry at that type of advertising promoting the sale of a neuro-toxin to the childhood age groups."

William Pardridge, MD inadvertently answered Dr. Elsas' question on excess phenylalanine. He stated, "Five servings of any NutraSweet product for a fifty-pound child exceed the FDA allowable daily intake recommendations. This intake would increase blood phenylalanine levels three hundred to four hundred percent."

Dr. Pardridge's research on aspartame focused on aspartame's influence on IQ, decision making capabilities, and changes in electroencephalograms (EEGs). He stressed the fact that eight thousand tons of aspartame were in the food supply as of 1986, producing eight million pounds of phenylalanine.

Senator Metzenbaum, chairman of the hearings, broke the venturesome mood when he placed a diet cola on a table in front of him, along with a pack of sugar-free chewing gum, a roll of hard candy containing NutraSweet, a can of diet powdered fruit drink, a container of laxative with aspartame, a jar of iced tea with NutraSweet, a box of sugar-free hot cocoa, and a carton of instant breakfast drink mix with NutraSweet. "The abundance of aspartame and the vagueness of its labelling are what I'm concerned about here, today. Especially in products marketed to children."

Comments of those gathered around the table indicated troubled feelings. Dr. Peter Dews spoke on behalf of the NutraSweet Company. When Senator Metzenbaum asked Dr. Dews for his opinion concerning the labelling of NutraSweet products, Dr. Dews stated, "I have not studied aspartame, but it's possible to confuse the issue by bringing out too much information."

I was shocked as I read Dr. Dews' statements. Dr. Dews never studied aspartame, yet he testified on behalf of its safety? Apparently, Dr. Dews had never published a study on NutraSweet, either. Nor had he studied the effects of amino acids or phenylalanine on children and on pregnancy. He also had never investigated aspartame in relation to seizures.

Jim Turner, JD, attorney for the Community Nutrition Institute of Washington, DC, who also spoke at the national symposium, objected to the FDA's procedure of aspartame analysis. Turner and Dr. John Olney began challenging G.D. Searle in the early 1970s concerning aspartame causing holes in the brains of Olney's laboratory mice. Turner had been tracking the aspartame saga from the very beginning. Turner stated, "There is a communication gap. The FDA's argument is there, but is not a scientifically conclusive connection to demonstrate a cause and effect relationship which they need to establish in order to take regulatory action. The point that the FDA is making is not that the events are not occurring. The point they are making is that they cannot prove they are occurring! One of the main points they make is that in examining seizure cases, they cannot rule out other possible causes of seizures. The very process they use, however, means they cannot rule out NutraSweet. It is important to understand that the FDA has not ruled out NutraSweet."

Turner asked the FDA to indicate any other food additive that has received more than fifty to sixty complaints. "In fact, this is more of an adverse reaction than what comes in the files on drugs," he stated.

As of 1988, it had been seven years since aspartame was approved as a food additive. Though initial safety concerns remained virtually buried under ruling postponements and continuous bungling of facts, the original aspartame issue prevailed. Negative health statistics directed to the substance were on the rise. Researchers had begun tallying increases in headaches, seizures, tumours, depression, and lesions. But the public would not hear about these statistics until many years later. Early aspartame critics came upon strong resistance. A "media black-out" was taking shape.

By 1989, aspartame was an ingredient in over two thousand products world-wide, and widely advertised in media and print. Nearly ten years later, that number increased to five thousand. Aspartame was now a multi-billion dollar commodity. Information against aspartame became more scarce.

In the isolated circumstances when these negative aspartame statistics surfaced, it is reported that advertisers threatened to pull sponsorship for those who supported anti-aspartame campaigns. According to United Press International, the "Aspartame Technical Committee," consisting of the NutraSweet company, Ajinomoto Company (G.D. Searle's Japanese supplier), several soft drink beverage companies and various food manufacturers using aspartame in their food products, allegedly launched a campaign to discourage researchers and critics from receiving Research Funding Awards Grants, especially those awarded by the International Life Sciences Institute, an organisation predominant in awarding grants to research scientists.

Samuel Molinary, co-chairman of the International Life Sciences Institute grant panel, accepted employment as G.D. Searle's director of scientific affairs. Shortly after that, Molinary accepted a position as Pepsico's research director.

The critics of aspartame never seemed to receive equal resources nor representation to complete sufficient studies opposing the use of the substance. Some independent researchers fell prey to repetitive grant rejections and strong administrative resistance.

Concerning this matter, Jim Turner responded, "Individual scientists cannot do their best and most objective work when they are financially beholden to their companies – companies who pay their salaries and who stand to make hundreds of millions, or even billions, of dollars based on the outcome of their research."

The NutraSweet Company spent millions of dollars building corporate research institutes, such as the Searle Research Centre at Duke University. They granted, and still grant, corporate scientists unsparing funds for ongoing aspartame research. It was at this time that the first national symposium on the safety of aspartame was held at the University of North Texas in Denton, November 8-9.

Sweet Poison

On January 2, 1992, the chief of clinical nutrition in the Assessment Section of the Clinical Nutrition Branch, HFF-265, at the Department of Health and Human Services compiled a memorandum documenting negative reactions to aspartame. The Department of Health and Human Services filed a report called "Adverse Reactions Associated with Aspartame Consumption," which lists a number of consumer complaints on aspartame filed by reported symptoms.

Headaches topped the list, with 1,487 registered complaints. (See Table at end of Chapter 9.)

The Public Health Service Memorandum also included a table on aspartame reactions registered by the product used. From a list of products, diet soft drinks registered 2,730 complaints and table-top sweetener (Equal) received 1,418 complaints.

After his reaction to aspartame, pilot George Leighton began investigating suspicious incidents related to flying. He wrote the following letter in April 1922 to the editor of 'US Air Force Flying Safety' magazine in response to an article they had published concerning Air Force pilots:

"I would particularly direct your attention to the potential altitude effects of methanol contained in aspartame. To my knowledge, there has been no investigation of its binding to the haemoglobin like carbon monoxide, thereby inducing hypoxia as suggested by Dr Phil Moskal. Perhaps your Air Force flight surgeons would be interested in pursuing this from a medical viewpoint. I do not have the resources to pursue such an investigation, and the FAA is stonewalling the whole aspartame issue. The FAA's inaction is very likely politically motivated. As a 'General Aviation News' article points out, Samuel Skinner, at the time boss of the FAA as Secretary of Transportation, was formerly employed by NutraSweet's law firm. Spotlight' magazine of April 6, 1992, revealed that his wife was also employed by that firm. Later, as President Bush's chief of Staff, Skinner was in an even more powerful position with direct influence over all government agencies, including the FDA, the FAA and your own Department of Defense. Through his past employment and his wife's employment with NutraSweet's law firm, it would seem that NutraSweet had a pipeline directly to the top."

He went on:

"I am continually appalled by the apparent indifference and inaction by various pilot-oriented organisations – FAA, Aircraft Owners and Pilots Association, the Airline Pilots Association, and others – to the in-flight hazard posed by pilots' ingestion of diet drinks or other drinks laced with aspartame, NutraSweet/Equal. In the 'Navy Physiology' article,for example, they stated: "Aspartame can increase the frequency of seizures... suscetibility to flicker vertigo or to flicker-induced epileptic activity." It means that ALL pilots are potential victims of sudden memory loss, dizziness during instrument flight, i.e., vertigo and gradual loss of vision. This certainly is a direct safety-of-flight item and should be dealt with as such.

"Instead, the Navy 'offer a heads-up to a potential problem.' Let me ask you a question. If you became aware of a component of every Air Force aircraft which was subject to sudden, catastrophic in-flight failure, would you simply write an innocuous 'head-up to pilots' at the end of an obscure article in your magazine? Of course not! You would take immediate emergency action to ground all aircraft until the safety-of-flight item was removed. Aspartame, NutraSweet/Equal, is that safety-of-flight item!

"I do not expect to see aspartame banned from the marketplace at this time; that is an unrealistic expectation. My immediate personal goal, however, is to have all pilots informed of the potential safety-of-flight hazard posed by aspartame. At least, they could then make an informed decision whether they wished to risk their lives and careers by playing airborne Russian roulette with aspartame-laced products. Of course, a more appropriate question may be asked: Should pilots even have this right when other people's lives are at stake?"

In May 1992, Dr. H.J. Roberts published his book, 'Sweet'ner Dearest'. In it, he illustrated the seriousness of aspartame's dangerous side effects using witty warnings of reality. Through the healing power of laughter, Roberts shared the experiences of many frustrated patients as he did at the national symposium.

Flying Safely magazine published an article in May 1992 stating that a single stick of gum with aspartame could cause negative symptoms in some individuals. The article explained that pilots who consume aspartame while in flight are susceptible to flicker vertigo, sudden memory loss and dizziness. The article went on to list the most common symptoms pilots suffered after consuming aspartame: seizures, headaches, dizziness, visual impairment, loss of equilibrium, tremors, retinal haemorrhaging, tunnel vision, panic attacks and depression.

Looking over all the information laid out in chronological order was much more shocking and disturbing than I anticipated. It was almost

ALIVE AND WELL

On my short break from researching aspartame and since my health had improved to the point where there was no sign of Graves' disease, I was inspired to reach for new physical challenges. In addition to devoting some quality time to my family, I decided to do some things I had always wanted to do. The first was signing up for a 150-mile bicycle race merely two years after my hospitalisation.

Having taught aerobics since the early 1980s, I was in good physical condition before I got sick. But the Graves' had set me back. I was weak and no longer able to lift heavy weights, bench press 120 pounds, or meet my average number of one hundred sit-ups a day. It took a year to build my fitness level back to where it had previously been.

I had slowly increased the bulk of my weights and had gradually added the number of sit-ups until I reached my normal one hundred again. I took walks every day and resumed my biking. I diligently increased my cycling to ten miles every other day during year two. In preparation for the race, I reached my goal of biking ten miles every day.

I looked forward to taking on this challenge. Bicycling 150 miles seemed sufficient proof that the doctor was wrong about my Graves' disease. I wanted to confirm to myself, once and for all, that my recovery was permanent and the Graves' disease would never come back. The race started from Midland, Texas. That day happened to be my thirty-seventh birthday, too. Boy, this is quite a benchmark for me, I thought as I stood in line to sign in. Its taken nearly two years to get to this point.

Sweet Poison

It wasn't happenstance that I was taking a quantum leap biking this race as I turned another year older. I was starting my life over. My extensive research into aspartame had been the first step towards the new me – at least on an intellectual level. Now the bike race would provide a fresh start for me physically. I'd never be the same person I was. Not after what I'd been through.

The entire race seemed impressively well organised, with comfort stations planned every ten miles or so. We were going to be able to stop if we wanted, chug a couple of bottles of cool water, and suck on a few orange slices that were to be provided for the racers. The event was going to be divided into two equal days of riding. We were to depart by seven o'clock both mornings and roll across the finish line somewhere around noon.

It wasn't long, as we started off the first morning, before the streets of Midland opened into the most beautiful scenery I had ever imagined. I eyed the breathtaking surroundings in awe. The sky was deep indigo with not a trace of a cloud in sight. Riding further into the desert, impressive geometric shapes of rose-coloured sandstone ascended like stone monsters; they seemed to rise out of nowhere. The great plateaux stood fixed against the azure backdrop like an oil painting positioned on a giant stone easel. The works of Georgia O'Keeffe – which I thought were beautiful – paled in comparison to the real thing that day.

"I've been across this desert many times before with the windows rolled up, the air-conditioner set on max, and the radio blaring. I never really noticed the raw beauty of this place before. Not like this, anyway," I shouted to another racer I was biking alongside. Biking through the desert paradise, I could smell sweet desert blooms, feel the sandy surface of the road slide beneath my pumping wheels, and actually hear the quiver of an eagle's wing as his six-foot stretch glided merely feet above me. At times, I thought he was following me. I hoped he was.

This was living! "Screw the Graves' disease," I yelled into the wind. I felt all the weight of my illness blow away as I pedalled faster and faster.

With seventeen miles remaining the second day of the race, I began my climb up and out of the West Texas Permian Basin. It was 110 degrees Fahrenheit on the road, and the climb uphill was tough work. I walked some of the way as sag wagons drove by pouring cold water over all the bikers. The inhabitants laughed wildly as they soaked us.

Although I was never a fan of having cold water poured down my shirt, that day it felt invigorating.

The exhilaration of finally reaching the top of the last hill reinforced my desire to triumph. Soaking wet, I remounted my bike and pedalled like hell. In less than two minutes, I was bone dry. I increased my pace. I wanted to finish the race having given it my all, not to mention staying ahead of the icy bucket handlers.

With only ten miles left until the finish line, I shifted gears and hauled. I thought about my Graves' disease, and why I joined this race. I pedalled faster. I gritted my teeth and said to myself, "I'm going to fly through that finish line, damn it! I'm going to do it!"

And, I did! I took first place on my team.

As I loaded my bike on the bus for the trip home, I gave it a pat of gratitude and a smile of warm appreciation. That bike helped me accomplish something I never thought I'd be able to do. A personal best, two years after being diagnosed with Graves' disease, I completed a 150-mile bike race – on my thirty-seventh birthday – through the West Texas desert – in July – and won!

The race convinced me that not only had I returned to health, but having been so close to death, I wanted to savour my children and my life. So, I did something else I'd always wanted to do: become a fire-fighter. In August, I joined the fire department. Still teaching at the university, I joined the department as a volunteer. I was lucky our fire station enlisted volunteers. This didn't mean I was any less of a fire-fighter. It meant I had the luxury to stay home if I needed to. My children thought I was wonderful. My husband thought I had lost my mind.

Even as a little girl, I played with fire trucks and dreamed about being a fireman. Growing up, whenever a fire truck passed me on the road, something inside me wanted to go, too. As my life continued to change, I decided it was time to live that dream. I'd never been more excited about anything in my life.

Finally, I felt the rush of riding in the fire truck; speeding to a fire clad in my bunker gear; sirens echoing through the streets; the anticipation of the emergency just minutes ahead; the strength and hard work required to save lives; helping people. I'm one of those people who stays cool in a crisis and thrives on assisting those in need. I love being a fire-fighter. I had thought my Graves' recovery and the bike race were challenging – but I soon learned that fire fighting was by far one of the hardest things I'd ever done. Every emergency demands complete mental and physical strength.

We fire-fighters kept our skills astute by training for several hours each week and every other weekend. We averaged at least three emergency calls every twenty-four hours. All fire-fighters wore a pager that beeped twenty-four hours a day whenever any emergency 911 call was reported in our area. My beeper impulsively bellowed day and night.

There was a unique bond among the fire-fighters. We saw and did things you couldn't imagine unless you witnessed them for yourself.

My beeper went off one Saturday at 2:15am. Beep Beep Beep Beep, arousing me from a deep sleep. Another emergency call. I cast the covers off and vaulted from bed. Lumbering my way through the dark bedroom to dress, I avoided turning on the light and waking my husband. I stumbled into the closet and stepped into my jump-suit one unsteady foot at a time, swished mouthwash around in my mouth, and immediately took off.

Leaving a quick note, I snatched my keys from the hook by the back door, slipped out of the house, and jumped into my Isuzu Trooper. I was fully awake by now, my adrenaline beginning to swell as I punched the accelerator and hightailed it to the station. In front of me and following behind, my friends and neighbours, who were also fire-fighters, responded to the call. I'm sure we all would have preferred to stay in bed, but there's something different about all of us. Fire fighting is a need, a challenge, a fulfilment, a duty, an undefined excitement.

There was a major accident on a country road south of town. By the time our attack truck and ambulance arrived on the scene, the neighbouring fire department had already begun cutting a woman from the car. Her car was smashed to one-half its width. The make and model were unrecognisable. As part of the fire crew, I approached the vehicle to assist in the woman's removal. She was so beautiful. Her long blond hair fell aside the stretcher as we gently placed her broken body across the mobile berth. I noticed she was wearing the same type of aerobic shoes I wore when I worked out. I wondered if she did aerobics, too. She was conscious when we arrived, but by the time the paramedics placed her on the stretcher, her heart was beating merely three beats per minute. The emergency helicopter was summoned and arrived within minutes. Its whirling blades rippled waves of sound that split the night as it gently set down in the middle of the highway like a winged dinosaur.

At the scene of most emergencies, there is so much activity

demanding multiple responses from all personnel that it is hard to stay focused on just one thing. That night, we fire-fighters thought we only had an accident scene to clean up until one of the paramedics cried out, "There is a baby seat in the back seat. Oh, Dear God! She has a baby with her."

Why would a baby be with her this time of night, we asked one another in horror at the possibilities.

I looked in the back seat of the crumpled car, but saw only an empty baby car seat. We began searching. Could the child have been thrown from the car? My thoughts immediately went to my boys when they were babies. Their faces flashed in my mind. Where was her child?

We searched inside and outside the car, all around the vehicle, in the ditches, and several yards on each side of the road. My fire boots projected gravel several feet ahead of me as I slowly inched forward step by step in my search. I walked to the top of the bridge where the woman's car had crashed and peered down onto the road below. I gasped. Something was lying down there.

Craning my neck for a better look, I slid down the embankment. Struggling to keep my fire boots from getting ahead of me, I got closer to what it was. Was it a child?

I rushed to the limp form and reached out only to discover with relief it was not the woman's child, but a stuffed toy. The trunk of her car was filled with toys and carpet and tile samples. The trunk had flown open upon impact, and, apparently, this toy had sailed over the bridge.

With a sigh of relief, I climbed back to the road and was immediately handed the fire hose. The car's gas tank was leaking. The smell of gasoline filled the air. While holding the hose, I noticed a pool of blood under the car door. I looked up to where she had been sitting. What a beautiful woman, I thought again. I kept seeing the paramedics lifting her onto the stretcher. Her long blond hair. Her aerobic shoes. She could have been my friend. I could have been her aerobics instructor. She could have been me.

Tidbits of information passed from one fire-fighter to another with compassion and concern. We pieced together the woman's history. She and her husband were building a house in our community, and while waiting for it to be completed, they lived in the apartments less than a quarter of a mile from the accident scene. She did have a small son, who was at home. "Momma's not coming home tonight. Momma's not coming home," one of the fire-fighters said as he picked up a crumpled piece of the fender lying in the middle of the road.

Her condition was rapidly deteriorating. There was no denying the truth. She was bleeding to death, drowning inside her own body. I could tell from the paramedics' faces how tough it was for them working on the young woman. There was blood everywhere, and they were losing her rapidly.

The police were now on the scene. More news filtered to us. She had been driving over ninety miles an hour when she lost control of her car. The car did two 360 degree turns airborne before it freakishly hit the side of the bridge guard rail exactly at her car door. Her entire body, internal organs and all, impacted at over one hundred miles per hour. When her car finally stopped several hundred feet from initial impact, she broke her air bag with such force she literally ripped the aorta from her heart.

It was only a matter of minutes more. Then she drowned in her own blood. She died before they could transfer her onto the Care Flight helicopter, though the medics desperately continued to work on her.

As the paramedics lifted her from the stretcher into the helicopter, her aerobic shoe fell from the stretcher. Then they turned and walked slowly toward the ambulance. They looked like cartoon characters flickering in the headlights from the chopper. I watched one of the paramedics stand in the middle of the road, staring at her shoe. After a few long seconds, he just walked away, leaving it there.

"What about her husband and little boy at home?" asked my partner, not really expecting a reply. We were all silent. We thought of her family.

"I don't envy the person making the phone call to tell her husband the news," I responded. We thought about our own families and, of course, our own mortality. After two-and-one-half hours, we finished cleaning up the debris strewn across the highway, rolled up the hoses, climbed on the fire engine, and left the scene in silence.

The devastating experience of being diagnosed with Graves' disease had affected me in so many different ways, small and large. With that as part of me and my experience as a fire-fighter – from the trauma of witnessing death to the invigorating challenge of fighting fires – I was reminded every day of the gift being alive really is.

Not only was I furious that I allowed myself to be suckered in by the "diet thing" to the point that I almost died, but I knew now I needed to get back to the basics of living.

Before my illness, my life was headed in the wrong direction. I was busy all the time. Too busy: fast foods, rush hour traffic, no time to relax, quick fixes with my family life. Recovering from a frightening disease and seeing the lovely young mother killed in her car awakened me as to how important it is to take the time to enjoy and appreciate life. I promised myself I would make vital changes; changes for the better. I'd swore I'd never compromise my health, my family, or my life again.

During one of our weekly fire training classes, one of the paramedics asked if any of the fire-fighters would be interested in going to paramedic school. "The department will pay for half of the school costs," he said to tempt us to sign up.

His plan worked. Over half the fire crew, including myself, consented to do it. The program was to take the better part of a year. Starting with six months of classroom study, it involved going on ride-alongs and preparing for the Texas State Board exam. Two months into the programme, the lesson was on I.O. inoculations, which are special techniques to stick an IV into a traumatised patient with failing blood pressure and veins impossible to stick with a traditional needle. In cases like these, the IV drip has to be administered through the calf bone, so the needle must pierce the bone. The instructor distributed turkey legs for us to practice driving needles through. That did it for me! I decided that paramedic school was more than I could handle – I couldn't successfully jab the long, thick needle through that leg bone without passing out. After that, I signed up to drive the ambulance for the department as long as I was allowed to remain in the front of the mobile hospital when we went on calls! My new duties as driver required a physical exam, so I made an appointment with the department's physician.

After a lifetime of filling out standard medical forms, you'd think I'd be used to filling out the routine pages of information, especially after itemising the stacks of forms required to deliver three children. Indeed, I thought I had become accustomed to the same meaningless medical questions on each form: family background, mother's medical history, father's medical history. Is heart disease in your family? Does diabetes run in your family? Ever since I was in elementary school, I was accustomed to filling in, "Unknown. Do not know family history. Adopted."

This time, I considered the mystery of my genetic background in a different light. I should know my medical history. I wanted to find out if my Graves' disease was inherited or not. If it was, it might have an effect on my children. I wanted to put my mind to rest about the Graves' disease once and for all.

So, I resolved to find my birth mother. I'd thought about doing this many times but more intensely since I was hospitalised with Graves'. In fact, since then I had been gathering bits of information about the whereabouts of my biological parents here and there, but I had not had the courage to follow through – until now.

I'd always been curious about my birth mother. I always wanted to learn the truth about my adoption, but I had been constrained by my loyalty to my adoptive parents. Now was different. I wanted to know if my birth mother had Graves' disease. This time, I finally stopped debating the assets and liabilities of finding out.

Growing up, my adopted parents shared what little they knew about my past. I know that immediately after my birth I had been placed for adoption in the Florida Children's Home Society. Then I was placed in foster homes from birth until six months of age. I developed chronic ear infections and endured horrible diaper rash during that period, so my adoption was put in limbo. However, after waiting for two years to adopt their second child, my adopted parents finally took me home, diaper rash and all.

I discussed opening my records with my adopted parents. "If it upsets you in any way, I won't do it," I promised. Mom and Dad were very supportive. They offered to help me any way they could. I really appreciated their attitude; I knew my decision to find my birth mother couldn't be easy for them.

Mom took my birth certificate out of the safe in the back room and handed it to me. "Here," she said, "this is yours. You keep it now." I studied my birth certificate. It had my adopted parents noted as my biological parents, and there was no time of birth recorded on the certificate. But I didn't let the lack of information stop me.

The next Monday, when university classes ended early, I got home before the kids got out of school. It was an ideal time to pick up the phone and call the adoption agency. I'd been putting this call off for too long. Just do it, Jan! Pick up the phone and call, I told myself. I dialled the Children's Home Society.

The line rang and a receptionist with a welcoming voice quickly answered. I began to explain my situation when she kindly interrupted, "I'll transfer you to someone who can help you."

She linked me up with a social worker named Betty Stoddard. My mind flashed to Mary Stoddard. I thought, Another acquaintance with the name Stoddard. A good luck omen? I hoped so.

Recounting my situation, I filled in the little I knew about my origin. Betty Stoddard confirmed that medical needs were indeed a reasonable foundation to open my adoption files. I was exhilarated!

After we finalised the initial order of business, she kindly spoke to me on a personal level, asking, "Are you sure you want to do this? Are you prepared to open doors that may not be easily shut? You may be disappointed in what you find," she added. "Do you think you can deal with whatever happens next?"

I was surprised at her warning. She sobered my enthusiasm by forcing me to look at the reality of what I was about to do. She was right. Once I opened the door, would I be able to shut it again? Would my birth mother?

If my mother had Graves' disease, would I eventually have to destroy my thyroid after all? And what would aspartame's role be then? Would I kind I devoted months, years even, to fighting something because of a lack of information about my family's medical history? However, if my mother didn't have thyroid problems, I could maintain confidence in my holistic healing and in my theory that aspartame indeed damaged my thyroid gland. I had come too far to turn away from the truth now.

"Thank you for caring, Betty. But, I need to go through with this to answer my medical questions once and for all. It's important for me to know my medical history. I need to know if my birth mother has any form of Graves' disease," I said with sincere appreciation for her concern.

As I placed the cordless phone back in its cradle, I took a deep breath and sighed. "Well, I'll see what's going to happen in the next few weeks. I'll just have to wait until I hear from Betty."

It was October when I heard from Betty – less than two weeks after our initial conversation. I knew something was wrong by the way she greeted me. She's found my birth mother and quickly, too, I thought. I knew my search had reached an end.

After a lifetime of speculation, I had the answers to my secret biological past. "Tell me what you found out," I urged, hurrying Betty beyond pleasantries. "Did you talk to her? Did you find out if she has thyroid problems?"

"Yes, I spoke with your mother, and she does not have thyroid problems," Betty said in a soft, slow tone. "Your birth mother has hypertension and Type II diabetes, which is adult-onset diabetes controlled by diet. Your maternal grandfather was healthy all his life until he died of a sudden heart attack at seventy-three years of age. Your maternal grandmother is seventy-nine years old and also suffers from hypertension. Your mother said her sister is healthy, and your stepbrothers and stepsisters are all healthy. No one has Graves' disease or thyroid problems of any kind."

Trying to stay cool and calm, I asked, "Did she say anything about my birth father? Can you find him, too?"

"Your birth mother offered no information regarding your biological father," Betty replied, her tone sad but calm. "When she entered the hospital to give birth to you, she listed him on your original birth certificate as 'Father Unknown.' She asked me to wish you luck with your illness, but to never contact her again. Her husband and children do not know about her surreptitious pregnancy in 1955. She will not risk uncloaking her secret."

Okay. Now what do I do? What do I say? I asked myself, trying to put my jumbled thoughts into words.

Pacing nervously back and forth across my home office with the phone to my ear, I felt a mixture of emotions at the outcome of my search. I could only imagine how difficult it was for Betty to relay such a message to me. I thought of the birth mother with whom, despite myself, I'd imagined a reunion. "This must have been a shock for my biological mother, getting such an unexpected phone call, I mean," I said as I sat down with a thump and continued unable to stop my bitterness. "I guess I ruined her day. The phone call she never expected to get. I wonder if she even knew if I was a boy or a girl."

Even though I was talking to Betty, I directed my comments to no-one. Calming a bit, I told Betty, "I was prepared for this outcome, really. I hoped it would work out differently, but I have what I need – my medical history."

Hesitating to even ask the question, I finally queried, "Betty, could you give my mother a message for me?"

Betty quickly interjected, "I can't contact your mother again, Jan. She asked me not to, and I must honour her request."

If I'd had more experience in situations like this, I would have delivered my message when Betty first contacted her. But now, it was after the fact. I had so much I wanted to tell her. I wanted her to know that I was not resentful at being adopted. That my adopted parents were great and I am fine. I wanted her to know she had three terrific grandsons. I wished I could tell her, You did the right thing in giving me up for adoption. I had a wonderful childhood and caring adoptive parents. But, Betty could not deliver my messages. My mother would never know anything about me. My one chance to impart a lifetime of thoughts was gone. I'll never have another opportunity like this again, I told myself.

"Betty, could you put a note in my file requesting my records be kept open in case my birth mother might want to contact me some day?" I asked softly as a lump formed in my throat and my voice began breaking up.

"Put in the file that she should harbour no regrets," said as I thanked Betty for her kind help.

Well, my search didn't have a fairy tale ending, but more importantly, I'd found out that my thyroid disease was not inherited. My biological past was no longer a mystery.

Now, I was completely convinced that aspartame had caused my Graves' disease. I knew I had made the right decision not to destroy my thyroid gland and to search for the real cause and a better cure for my illness.

KATRINA

O ne Saturday morning as I sat reading the newspaper, the loud ringing of the kitchen phone broke my concentration. Setting my coffee on the kitchen table, I rushed to grab the cordless, hoping the ringing wouldn't wake the kids.

"Hello?"

"Jan?" a voice timidly asked. "I'm sorry to call so early. Did I wake you? This is Mary Stoddard."

"No, no," I replied. "I'm up reading the paper. What's on your mind?"

"I'm sorry to bother you," Mary apologised again, "but, could you possibly come over today around one o'clock?"

Without thinking about the day's schedule, I quickly answered, "Sure. What's going on?"

"Well, I'll tell you when you get here," Mary replied mysteriously. "I'd like you to meet someone special."

"Okay. I'll see you around one."

As I hung up the phone, I thought about the day at hand. I hoped Chuck could handle Sean's TaeKwonDo tournament, Brian's soccer game, and Alex's baseball practice without my help. Even though I hadn't talked to her in a while, when Mary, who has done so much for me, ever called and asked for a favour, I couldn't possibly refuse. Anyway, the sound of her voice had piqued my curiosity and I was sure it might have something to do with aspartame.

It was not unusual for Mary to deliver a vague message over the phone, either. Mary remained cautious when discussing aspartame issues over the wires. Pioneering the Aspartame Consumer Safety Network since 1987, she suspected she had been monitored for years. With an hour drive to Mary's ahead of me, I started breakfast for the

165

boys and headed for the shower.

I arrived at Mary's before the appointed hour. Curious about this meeting, I eagerly walked to the door. Mary answered my first ring of her doorbell and ushered me inside.

"Janet Hull Smith," Mary said formally, "I have someone I want you to meet."

Much to my surprise, a precious five-year-old girl with a big red ribbon bouncing loosely upon her head skipped over to me and tightly wrapped her arms around my legs. Radiating a contagious smile, she looked up at me, a perfect stranger, and said, "Hi! I'm Katrina!"

Mary, Katrina and I went into the living room where I met Katrina's parents, Jay and Carmen Carradine. We sat down together, and the mystery soon began to unfold as her parents related Katrina's terrifying story.

Katrina had been crying and complaining that her ear hurt terribly the very day she went for her three-year-old check up in Indiana on Wednesday, January 5, 1994. Her paediatrician checked her ear. "Well, Mrs. Carradine," the paediatrician stated as she tore a sheet of paper from a rubber-lined pad. "Katrina appears to have an ear infection. Here's a prescription for an antibiotic."

"Katrina has had ear infections before," replied Katrina's mother, Carmen. "But, she's never complained of so much pain. Why this time?"

"I really can't answer that question, Mrs. Carradine," the doctor said looking unconcerned. "I'm sure it's nothing to get worried about."

Carmen pressed on. "Odd things have been happening to Katrina lately. She has become clumsy, particularly compared to other children her age. Sometimes, she acts as if she's blind, literally running into things. Occasionally, she falls." Carmen added, sharing her concerns with the doctor. She continued, "and Katrina has complained of a stomach-ache almost daily over the past six to eight weeks."

The doctor suggested Katrina see a neurologist regarding her blind falls. She added, sounding more concerned, "If she doesn't improve in a few days, bring Katrina back in, and we'll run some tests."

Carmen was worried.

Frequently during the past months, Katrina seemed hyperactive at times, and at other times she appeared sleepy and sometimes slurred her speech. She never acted in a reliable manner. Her mother wondered if this was typical three-year-old behaviour.

Katrina commonly had loose stools, but over the past several months she also experienced diarrhoea and cramping. Two days after the doctor's appointment, Katrina's ear pain worsened. She continued falling, once bumping the back of her head quite severely. She kept complaining that her head hurt where her ear was infected. Carmen called the paediatrician, who guessed the antibiotic must not be working. She prescribed Vantin, an antibiotic she had used before, and Tylenol with codeine. Katrina vomited several times that night. The Tylenol never had a chance to get into her system because she couldn't keep anything down. Her parents waited until morning to start the Vantin and tried a second dose of the Tylenol.

Carmen was not convinced Katrina simply had an ear infection. She believed this time her ear was sore because it was swollen and red, not the other way around.

By Saturday, Katrina was completely lethargic and still feeling nauseous. Carmen held her most of the day when she was not sleeping. Katrina kept saying she was not hungry. Finally, her mother managed to get her to eat some sugar-free yoghurt. Katrina liked yoghurt and Carmen hoped she would keep it down. However, the little girl vomited several times that day. To cool her throat and settle her stomach, her mother gave her some cold diet cola which she sipped.

The following day, Katrina was even more unresponsive and didn't want to get out of bed. Her parents decided to take her back to the doctor the next morning if she did not improve. At 1.15pm, Carmen had to leave for work. Katrina wouldn't get up for lunch as she normally did. Rather than disturb her sick child, Carmen let her sleep. She almost left without checking on her, but changed her mind at the last minute. Carmen entered the little girl's bedroom and found Katrina lying on her side with her back to the door. As Carmen rounded the bed to check on Katrina, she saw her little girl was staring strangely at the wall. Carmen gently called her name, but Katrina did not respond. Confused, Carmen said a little louder, "Katrina?" Still she did not move. Alarmed, Carmen called Katrina's father, Jay, who had just sat down to give a bottle to their three-month-old baby.

"Jay," she yelled in panic. "I think something is wrong!"

She turned Katrina on her back. The sight was one she would never forget. Three-year-old Katrina lay unconscious with her eyes wide open and glaring. Her lips and fingernails were blue. The right side of Katrina's face was covered with mucous and saliva from a puddle on her pillow.

Sweet Poison

Jay rushed to Katrina's side and looked at his daughter. He cried out, "She's not breathing, Carmen!"

Carmen picked Katrina up. As if in slow motion, they discussed what to do next. What would they do with the baby? Where would they lay Katrina? "Carmen, call 911!" Jay screamed, and began to give his daughter CPR.

Within one minute, two policemen ran through the front door without a knock. A minute later, an ambulance and a fire truck screamed to a stop outside the house. There were people everywhere asking questions. Carmen was still on the line telling the 911 operator what was happening. Although Katrina had started breathing again, the paramedics saw that her left side was totally paralysed. They asked Jay and Carmen if they had any idea why. The parents shook their heads. No one knew what was wrong with Katrina.

While being transported in the ambulance Katrina had a seizure and stopped breathing again. In the emergency room, Katrina kept crying out, "Mommy, Mommy." Carmen's heart was breaking.

Katrina sat up in her temporary hospital bed, arms extended for Carmen to hold her. Carmen held her tightly, fighting back tears that would have frightened Katrina. As Carmen set her back down, Katrina slipped into unconsciousness again and had to be resuscitated a third time. Carmen was asked to leave as the medical staff intubated Katrina. Jay finally arrived after dropping the baby off at a friend's house. Carmen filled him in on the situation. Katrina was given Valium, Dilantin, and Versid. The hospital performed a CAT scan. After four hours, Katrina was transferred to Riley Children's Hospital at Indiana University.

During the weeks that followed, while their baby continued to live with friends, Jay and Carmen received an incredible education.

Doctors performed an MRI without a contrast, but saw nothing. They did a spinal tap. Nothing. They thought they had her stabilised with anti-convulsants. They extubated her the next morning, moving her to the toddlers' unit.

That night, she complained of another bad headache. She could get no relief.

However, not long afterward, Katrina started acting strangely. Her hands began to twitch. Carmen held her while she called for the nurse. The nurse arrived, but just stood in place watching Carmen hold Katrina. Carmen grew angry. "Why don't you do something?" she questioned. She found out later that the nurse was not acting indifferent, but was watching and timing Katrina's seizures. There was nothing more she could have done. She was unaware that Carmen had never seen Katrina have a seizure before.

Within fifteen minutes, Katrina had another seizure, this time with more arm and mouth movement. Again, fifteen minutes later, another seizure overcame the child. This continued until 7:30am. Back in the ICU, Katrina slipped into status epilepticus. She stopped breathing again. They reintubated her and performed an MRI with a contrast. This time, they finally saw something. No one knew what it meant, however.

Katrina was put into a pentobarbital coma for one week to stop the status epilepticus. Otherwise, Katrina would have died from the damage to her brain by the constant seizing. The doctors wanted her brain wave to be as flat as possible, referred to as "burst suppression." During that time, Katrina was on complete life support and a constant EEG monitor. She had an arterial line in her ankle to draw blood hourly so the doctors could check for blood gases and drug levels.

She required a blood transfusion to replace what was being continuously drawn. She had two central lines: one in her neck and the other in her groin. Each central line had two lines leading into it. One was for feeding, the others were for administering various drugs. Her urine was collected hourly.

Then they put Katrina into isolation as they suspected she was contagious. Jay and Carmen were required to wear masks to stay with her. What a horrible sight it was for her parents to see. Their daughter was laying lifeless before them. It was a long, long time before she would come back. Jay and Carmen wanted answers! They asked the doctors why their daughter was so sick. What had caused this sudden illness? They complained they were being kept in the dark. No one offered them any explanations. As if they were disruptive children, they were scolded, "Look, we have a live child to work with here. She could easily be dead. You're lucky. Don't complain. All you can do for Katrina is support her. You'll just have to wait and see what happens." At that time, the Carradines were informed that sixty percent of the children that came to Riley for a diagnosis never get one. "So, don't get your hopes up."

For the two weeks while she was in a coma and on life support, Katrina's family didn't know whether she would live or die. If she did live, could they keep this from happening again? No one could give them any answers.

Sweet Poison

When Carmen was not at the hospital, she was searching for the answers. She went to the medical library, and continuously read articles, books, and research to try to help her daughter. Like the doctors, she found nothing.

Jay drove to Chicago late one night to have a friend run a test on some of Katrina's blood. He thought she might have lead poisoning. Nothing. They had their furnace checked for carbon monoxide leaks. They checked her blood for that, too. Nothing.

Chocolate. A friend mentioned an article she had recently read on theobromine, an ingredient in chocolate. She knew Katrina had an insatiable, almost embarrassing, craving for candy. Another friend suggested a possible carnitine deficiency. Still nothing.

Test upon test was run on Katrina. She had a CAT scan, three MRIs, three spinal taps, and so many EEGs that Jay and Carmen lost track of them all. The cost was astronomical. Katrina soon developed problems with her liver. A liver biopsy was performed. A muscle tissue biopsy was also done. An ophthalmologist was called in to check for a Kayser-Fleischer ring indicating Wilson's disease. Nothing.

A cytogenetic test was performed, an abdominal ultrasound of her liver and gall bladder. A Visual Evoked Response test and Brainstorm Visual Evoked Response test were performed. Nothing. The list went on and on and on. Nothing.

An infectious disease specialist diagnosed Katrina with non-infectious encephalitis. Her paediatric pulmonologists agreed she had encephalitis, but believed it to be viral. Three spinal taps did not support that opinion, however.

Katrina was then examined by a gastroenterologist and a neurologist. The neurologist's final impression on April 15 was, "Status-post status epilepticus of unknown etiology with evidence of a meningo encephalitis on MRI scan."

After Katrina came out of the coma, Carmen rocked her, talked to her, read to her, played music for her, and waited days upon days for her to "come back" from the deep sedation.

The time at the hospital spent waiting for Katrina to get better flew by quickly thanks to the constant flow of friends who came to keep Jay and Carmen occupied. No one could offer answers, only support and love.

However, questions continued to haunt the Carradines. Who knew the answer to why Katrina was suddenly fatally ill? What caused this to happen in the first place? Katrina was always healthy. Other than ear infections and the unusual symptoms over the two-and-a-half months leading up to her hospitalisation, she was a model baby. Who or what was to blame?

Jay and Carmen pondered the fate of their child. What should they do next? What would be the fate of their second child?

Katrina was finally released from the hospital after eight weeks. The bill was three hundred thousand dollars and yet nobody knew what was wrong. Her parents brought her home with little to go on. They didn't know what caused her near-death experience or if it would occur again. The doctors never determined what really happened.

The doctors wanted Katrina to remain on anti-convulsants for lack of anything else to do. After being on life support for two weeks, six weeks at Riley Children's Hospital, being transferred to a rehabilitation hospital for another two weeks, Katrina's illness was still shrouded in mystery.

Six weeks of tests, including three lumbar punctures, constant EEGs, a liver biopsy, a muscle tissue biopsy, echocardiography, genetic screenings, abdominal ultrasounds of an enlarged liver and high enzyme readings, four MRIs, tests for Wilson's disease, herpes, and hepatitis, hourly blood tests, lead poisoning tests, carbon monoxide poisoning tests, cytogenetic tests and countless others the Carradines didn't even know about, revealed nothing. An infectious disease specialist saw Katrina for a while, but found nothing.

She remained hospitalised for six weeks; on life support for two weeks, and in ICU for four weeks.

Katrina's final diagnosis remained meningo encephalitis of unknown etiology. No virus was found, no bacterial infection discovered. The doctors talked of a possible toxin, but it was never pursued.

Katrina finally went home. But, her parents asked themselves, for how long?

Six months later the Carradines moved to Dallas. A new friend and neighbour suggested Katrina's symptoms were the result of a reaction to aspartame. She gave Carmen a pamphlet published by the Aspartame Consumer Safety Network. Handing her Mary Stoddard's number at the ACSN headquarters, she said, "I know you're still agonising over what caused Katrina's illness and it couldn't hurt to check it out."

As Jay and Carmen read the pamphlet, for the first time since Katrina's illness, they found some answers that made sense. Those answers lay with the diet they had unknowingly fed their three-year-old thinking it was safe. Katrina liked powdered drink mixes. They contained aspartame. Carmen fed her diet yoghurt. It too had aspartame. Katrina took sips of Jay's diet drinks. Carmen used aspartame products during pregnancy to keep her weight down. Her OB/GYN said it was fine. These foods were given to Katrina occasionally, maybe even several times a week. But could she have taken enough aspartame to create such serious symptoms? It didn't seem likely.

Carmen called the NutraSweet Company explaining the situation and asked if they had ever received complaints like hers. They told her they had no idea what she was talking about. They mailed her brochures advertising all the products NutraSweet is found in, claiming complete safety.

The information they sent provided the last answers that were unintended. Carmen found out Katrina's daily children's multiple vitamins contained aspartame. Carmen never knew they contained the substance because the chewable vitamins were not labelled "sugar-free."

Carmen's mind began spinning back. She had bought the popular children's vitamin for Katrina in October and religiously given them to her daughter until January. On October 13, they had to take Katrina to the emergency room because she had stumbled and fallen. There were strange side effects. Katrina had a hard time standing. She'd stand in one spot with her finger pointing out, turning in circles and mumbling. Carmen talked to the paediatrician about Katrina's strange reflexes. To test Carmen's concerns, the paediatrician asked Katrina who her mommy was. Katrina went over to the nurse, pointed, and called her "Mommy".

Then, Katrina could no longer see well. She began to develop frequent earaches. Her mother started Katrina on a children's pain reliever, unaware it also contained aspartame. Katrina was given vitamins and pain relievers containing aspartame without her mother's knowledge. And the cycle continued: more vitamins, more aspartame, more earaches, more aspartame-laden pain relievers.

Home from the hospital, Katrina was placed on prescribed anti-convulsants. Carmen did not start her back on vitamins until she was seizure-free for over a year and weaned from the medication. Then Carmen started Katrina back on her favourite chewables. One week later, to her parents' horror, the symptoms of her illness reappeared. Katrina began complaining of stomach pains and diarrhoea, and once again the little girl began stumbling and falling down.

It was about this time that Carmen received the information on aspartame. She began to put two and two together.

Katrina had suffered the toxic effects of aspartame poisoning: acute toxicity, lethargy, confusion, impairment of circulation, severe headaches, abdominal pain, vertigo and temporary vision loss. Katrina also had nausea, unsteady gait and unusually high liver enzyme levels.

Katrina had already suffered so much and had a hard life ahead of her. She was in need of speech therapy. Her behaviour was unpredictable. She required special schooling. Her diet would have to be watched closely from then on, including intense scrutiny of all her meals and snacks. Eating out was risky; who knew what additives would be found in foods served in restaurants?

Katrina would have to be very careful since aspartame is hidden in so many things. Her life would never be the same.

Despite these handicaps with which Katrina was left, Carmen and Jay were very grateful to finally learn what was poisoning their daughter. Carmen wrote the following letter to Mary Nash Stoddard at the Aspartame Consumer Safety Network to express her thanks:

"Dear Mary, After talking with you, I felt a kind of excitement and relief because I was finally getting an answer to what happened to my daughter. Just ten days after her third birthday she nearly died, and would have had, my husband not known CPR, and had an ambulance and medical team not been only two blocks away...

"Is my child's life worth nothing? It makes me very, very angry that the FDA has allowed something to be put into my children's food that can harm them and even take their lives. The FDA knows it has already taken lives. Are millions of dollars being spent buying up books and other literature so the public doesn't have a chance to read the truth?

"I am very grateful to have been informed, and I know from our prior experience that my child could be dead by now if the ACSN had not chosen to speak out. Katrina had so many symptoms listed on the FDA complaint list, I knew she was a victim of aspartame poisoning. I knew it, even before I figured out where she was getting her largest regular dose of it. It was helpful for me to be able to put down the words, kind of like therapy. I will remain in contact and will do what I can to help you in your efforts to let people know the dangers of aspartame. Thank you for your help, and advice, and for listening!

Yours very truly, Carmen Carradine."

BACK TO BASICS

Between the discussion with Betty Stoddard about my biological mother and meeting Katrina, I wasn't able to sleep much over the following weeks. One evening, however, fatigue caught up with me, and I fell asleep early.

Unfortunately, I was awakened around 1.30am. My dog was very restless, pacing back and forth by the front window. She was whining, yet wouldn't go outside when I opened the door. I peered out to investigate what was upsetting her. Another dog was in the yard. It looked like a large German shepherd, but there was something odd about this dog. Upon blurry examination, I realised it was a coyote. A coyote? In the city? For some reason, that didn't scare me.

Enthusiastically, I waited for the coyote the following night, but it didn't appear. Native Americans would say, "It was a good omen to have a coyote come to you." But in this day and age in Dallas, Texas, I realised the animal was probably lost and hungry, so I warned my neighbours to watch their cats.

Around 1.00am a few nights later, I was awakened again, but not by the coyote whom I never saw again. This time, it was an EMS call on my beeper. Apparently, while driving on the highway around midnight, a couple rammed into the rear of an eighteen-wheeler. The impact sent their car out of control and wrapped it around a tree, which was where we found them. Before the crash, their four-year-old daughter had been asleep stretched out in the back seat. No seat belt, of course. Upon impact the little girl was thrown around the car like a racquet ball bouncing against wooden walls.

Dressing quickly, I hurried to the scene. We removed the victims from the crumpled vehicle, and the little girl was flown in critical condition to Ft. Worth Hospital on the Care Flight helicopter. Our crew transported her parents by ambulance to the neighbouring hospital.

The two different events – the natural beauty of the misplaced coyote and the trauma of that family's accident – now imprinted themselves on my mind and spirit. I felt like they were a call-to-arms.

I knew the "goods and bads" of big city living. Now I wondered why I lived amid the confusion, endless traffic, pollution, and crime continuously weaving in and out of urban life like a silent serpent looking for unsuspecting prey.

I've always loved being outdoors, which was one reason why I majored in the Earth Sciences as an undergraduate at the University of Texas at Austin. I crafted nature into my career plans. I held visions of working as a park ranger somewhere in the Colorado back country. When an excellent job opportunity tempted me back to Dallas the year after I graduated, I signed on with the Sun Oil Energy Development Company as a uranium analyst. I returned to Dallas with the intentions of staying temporarily.

However, the opportunities for advancement kept coming and I accepted each one. Before I knew it, I got married, bought a house, and started a family. I sank deeper and deeper into the "big city" standard of living I had always wanted to avoid.

When I became sick with Graves' disease, I was one of the many fake food, fast food junkies stressed out on city life. Despite all I suffered, I am thankful to the disease for being the divine hand of reason whose blow made me see reality. I now know I had been lured into a lifestyle that was killing me.

As a child growing up in Dallas, I spent hours looking out my bedroom window and staring at the beautiful landscape that surrounded our house. I would gaze at sleek horses and magnificent longhorns grazing amid vast pastures of grass and painted wildflowers. Comical roadrunners shot across our driveway unannounced. Woods and native songbirds were everywhere. The creeks were filled with crawdads, small lobster-like creatures the boys used to put down our backs on the playground at elementary school.

Within ten years, I witnessed Dallas explode in magnitude. The ranch outside my window was sold. The view now became one of stores, gas stations, housing developments, vapour light posts and neon shopping signs. I first learned to drive in city traffic and rented my apartment down a crowded street from my old high school.

This was urban reality. It's part of what I believe is steering people out of balance these days. Big cities are exciting but they are also the breeding ground for fast lives and foods, dieting frenzies, and eating disorders. They are the demand centres for convenience. Where is nature's place in the growth of contemporary American cities? I want to know. Where are the farms and fresh foods? How healthy are we city dwellers, really?

Nature invariably suffers. To support a growing population, trees have to be cut down to provide spaces for buildings, the wildlife are killed or run away, and thick layers of concrete spread over old farms to provide roads, neighbourhoods, and shopping complexes. Nature, along with the basics of living, keeps getting pushed farther and farther away.

To most people, pollution refers to emissions from cars, factories, and airports. To an environmental engineer like myself, pollution represents water and soil contamination from unnatural sources that is unable to dissolve or break down. I have come to realise it is time to include chemical poisoning from food as pollution, too! Getting sick from pesticides and chemicals is no longer the exclusive health danger. Chemicals added to our food supply as "food additives" cause disease in people, like myself. Until I dealt with the pollution of my own body, I never realised that pollution can also occur within us.

It was time for me to get back to a simpler life before I got so far out of control that there was no return to normality. I shifted from my hectic Dallas lifestyle to a slower, more natural pace. It was the only way to completely heal from Graves' disease and become truly healthy. I had allowed my big-city lifestyle to turn me into something unnatural, and I wound up fighting for my life because of it.

I came to love a book called 'Women Who Run with the Wolves' by Clarissa Pinkola Estes, Ph.D. One of my favourite passages describes my new outlook and ideals so well:

"We know that for creatures to live on Earth, they must at least from time to time have a home place where they feel both protected and free…

One will awaken to the voice calling from home, back to the core self where one's immediate wisdom is whole and accessible. From there, a woman can decide with clear seeing what it is she must have."

Our surroundings are powerful influences. I believe people are forgetting just how forceful their environments can be. All living beings fit into certain spaces like the different wooden shapes on a pegboard. Some people try to fit into spaces they aren't "cut out" for.

Others slide into the precise spots they belong. If people become more sensitive to their environments, they will be able to learn how to find the right surroundings best suited for their health.

Experiencing the debilitating symptoms of Graves' disease showed me that I had been totally unaware of what my hectic lifestyle was doing to my health, both mentally and physically. Although it took some time, study, effort and reflection, I became more aware of my diet and my environment. However, one shouldn't have to become deathly ill to return to the basics of good living.

After revamping my family's diet, one of the first things I did to simplify my life was to convince my husband to sell the house in which we lived and move. My husband was a self-employed contractor, so changing locations was fine with him. We moved to a small quiet town outside Dallas where we didn't have to lock our doors and people knew one another. We looked out on white picket fences, green trees, and peace and quiet. It was safe to take a walk after dark, and the schools took a personal interest in my children's welfare. We all felt a great sense of community and belonging.

I continued teaching, but decided it was an ideal time to return to school myself to get a graduate degree and professional certification in nutrition. I still kept in touch with Steve Fugua, putting into practice all he had taught me about nutrition. I also kept up my relationship with Mary Stoddard, helping her sift through continuous case studies and the endless volume of letters and documents she received every day from people who were having bad reactions from eating products containing aspartame. She continuously relayed information from publications and books. She told me about the book by Barbara Mullarkey who had recently left the Wednesday Journal, titled 'Diet Delusion'. In the book, Mullarkey wrote that FDA records showed 903 seizures and convulsions, 1,487 headaches, and four deaths associated with aspartame. I kept abreast of the ongoing events occurring in the controversy, like statistics on seizures and pilot issues, and all the latest research available.

I still thought back to my illness and how I had suffered and almost destroyed my thyroid gland. But now I was in perfect health, feeling great every day. Things were going so well that I decided to open a fitness studio in town. I worked out every day, anyway, and Sean, my oldest, was in TaeKwonDo. So, now to my experience as an environmental engineer, college professor, and aerobic teacher and fitness instructor trainer, I decided to add fitness studio manager. I asked my friends and neighbours what they thought about my idea; they all supported me and promised they'd join.

So, four years after my recovery from Graves' disease, I opened a fitness studio and nutrition centre. I now worked out twice a day, teaching low-impact aerobics, step aerobics, and weight training. The TaeKwonDo programme moved their classes into my studio, where kids and adults in white-belted uniforms kicked and chopped invisible attackers when the floor wasn't being used for aerobics. I also offered nutritional counselling and sold vitamin supplements. I began my intensive studies at the National Institute of Nutritional Education (NINE), so I could receive my Certified Nutritionist degree (CN) at the end of two years and become more involved professionally in the health and fitness industry.

The idea of becoming a Certified Nutritionist came to me partly because, during my recovery, I discovered there were different components to proper nutrition and that healing from a disease was like fitting pieces of a puzzle together, one by one. Even before I enrolled at the NINE, I had researched and studied nutrition and related issues. Now I began working towards my professional certification. In the course of my study, I developed a plan for better nutrition and a healthier life for myself and my family, and I hoped for numerous future clients. I want to share these precepts with you:

First, purify your environment as much as possible. Any toxins from both inside and outside the body should be removed. Inspect your house. Specifically, houses and places of employment should be inspected for old pipes, leaking insulation, unsafe drinking water, dusty air vents, pesticides, radon and the like.

As for diet, eat as many chemical-free, natural foods as possible, and minimise buying processed and packaged foods. Metals, such as aluminium and nickel, along with a variety of toxic chemicals like methanol, are used in the manufacturing of foods. No one should eat these substances. The manufacturing and packaging of foods removes natural vitamins and minerals too. Many times, artificial fillers are injected into foods to stretch profits. So, it's best to eat fresh and pure and avoid the fake stuff. Also, it's important to drink bottled or filtered water, and lots of it.

Exercise daily. Exercise keeps the blood flowing, and blood carries nutrients and oxygen to the entire body.

Millions of Americans are on reducing diets every day. Often they are involved in artificial "diets" and diet programs. However, if you lose one pound of weight by starving yourself and eating "fake" foods, you actually lose approximately three-quarters of a pound of fat and one-quarter pound of lean muscle mass. Yet, if you lose one pound of body weight following a whole foods diet with regular exercise, you actually lose approximately one and one-quarter pounds of fat and gain one-quarter pound of lean muscle. That's why I believe people who are trying to lose weight shouldn't diet at all. Instead, they should try to reduce by eating natural foods and exercising regularly. There's no such thing as healthy dieting, but there is such a thing as a healthy lifestyle, which includes balancing your life and purifying your environment.

Sadly, most people in large urban areas rely more and more on fast foods and foods fused with high degrees of preservatives for their main sources of nutrition. I believe this unhealthy lifestyle may be killing them.

The keyword is "over-indulge." People don't have to completely give up fast food and convenience foods altogether, but they must be aware of how much and how often they are indulging in them and try to minimise the amount of chemicals they are eating.

When I go to the grocery store on weekends, sometimes I feel like I'm on a Texas cattle drive. The stores are filled with jittery, anxious people who are always in a hurry. The checkout lines are long and slow.

Most shoppers, especially those who have a hectic, harried lifestyle, buy frozen dinner-type products, canned diet drinks by the case, one or two bakery items, several packages of low-fat this and fat-free that and not much else. I rarely see fresh fruits and vegetables, fish and chicken, dairy items, deli breads, nuts and not-from-concentrate juices in shopping carts.

Don't be fooled. The most natural foods are the healthiest. The easiest way to be sure what is in your processed food is to read the nutritional information on the label. Most are reduced calorie or fat-free. What they should be looking for is the natural content with the least chemical ingredients. It's hard to find packaged foods with fewer than ten additional man-made ingredients. I have a half serious rule that if my children and I can't pronounce the ingredients, I don't buy the product. The purer the food, the fewer the ingredients. And remember, there are no labels on fresh foods because what you see is what you get.

Of course for those with crammed schedules, shopping frequently is difficult yet it is important. Real food will spoil more quickly than manufactured food and should be eaten shortly after it's purchased. The best time to buy food is while it's bursting with live nutrients. Eat as much fresh food as you can. Active food enzymes and nutritional "jewels" don't survive the fabrication process very well. They're often destroyed somewhere between the boiling, freezing and vacuum packaging.

After my horrific experience with aspartame, I have learned to avoid processed, artificial, counterfeit, sugar-free, fat-free, calorie-free, responsibility-free foods. I only eat real food. It's more expensive and it takes more time to prepare, but it's healthier and could save your life. Besides, with the money saved on health care down the road, it's worth it.

I frequently consulted Steve Fúgua when I had questions or problems or needed his years of experience to help me with my studies. My friend and mentor, Steve taught me a great deal right from the start. I remember when I first met with him years before, he recommended I have a hair analysis done. "What is a hair analysis?" I had asked.

"A hair analysis is the highest calibre of laboratory science a nutritionist can use," he had answered. "This is something everyone should have done."

Steve had led me into his office and pulled out a lofty stack of papers. He had gone on to explain, "A hair analysis is a very technical laboratory analysis assisting physicians and health care professionals, clinical nutrition specialists, and biochemists in more than forty-five countries. It provides the nutritional data necessary to aid in the treatment of patients, and supplies quality data for research."

"I wish my doctor had been open to something like this when I was in the hospital," I had told Steve.

I had a hair analysis done that evening. Now I recommend to all my clients, friends and family, as well as those I talk to at speaking engagements, that they have at least one hair analysis in their lifetime. Steve had explained that the hair holds an imprint of all vitamin and mineral levels in the body and reflects minute levels of toxins deposited in the tissues. As people become more polluted, a hair analysis can detect specific toxins that are stored in the body. It's a valuable diagnostic tool.

Less than two weeks later, the results of my hair analysis had arrived. "Your hair shows me that the specific mineral and nutrient levels depleted by your Graves' disease are still low – most likely due to the toxic chemicals from the aspartame that totally saturated your body several years ago," Steve had said as he read the results of the analysis. "Your chromium, zinc, and vanadium levels still need boosting. These are the essential elements responsible for thyroid function. We need to keep building these levels until all the damage is restored. It may take a few months to a year depending on the degree of damage. We'll do another hair analysis in six months or so. We will be able to tell then if we have completely repaired the gland."

"Your pH level is much too alkaline, too," Steve had continued. "The more alkaline you are, the closer to bad health and death you become. As a matter of fact, when you die, your body is one hundred percent alkaline. You are a geologist, Jan, you know that fossils are pure alkaline. Lifeless, dead food products are pure alkaline, too."

I now know that after more than thirty years of eating alkaline foods, like bleached white flour and margarine, It's not surprising that I'm too alkaline. I remember asking him what I could do about it.

"It's important to maintain a high acid level in the stomach in order to digest your food properly," he had answered. "Good digestion is the key to good health."

With Steve's guidance, I had learned how to adjust my diet. After meals, I began supplementing my diet with a digestive enzyme rich in papaya, pepsin, betaine hydrochloride, and pancreatic enzymes. I also began eating the raw lemon that's balanced on the side of my glass of water or iced tea, and I enjoyed a glass of rich red wine to aid my digestion.

I knew that if you have stomach problems or worry about too much stomach acid, you should eat raw cabbage at least three times a week, because cabbage rebuilds the stomach lining. I picked up this trick from my studies on Russia during my years teaching at the University of North Texas. Steve, had known about it for decades. "Cabbage contains mucin, and the stomach wall is made of mucin," Steve had explained. "The mucin from the cabbage keeps the lining of the stomach wall thick and healthy."

Steve had added, "Research has proven some stomach ulcers completely disappear when raw cabbage is eaten at least three times a week. In fact, I have clients whose bleeding ulcers have disappeared after eating cabbage every day for merely two weeks."

Since those frequent consultations with Steve and with my ongoing studies, I continued developing what I believed was a healthy lifetime nutritional plan. I concluded that each meal should consist of seventy-five per cent raw foods. These include high fibre foods and grains, lean meats, fruits and their seeds, and an abundance of spring or filtered water. However, it's difficult to get all the nutrients and minerals that you need from your food. So, every day I carefully maintain a regular supplementary programme.

In the early days of my recovery from Graves', I supplemented the following vitamins and minerals: natural chromium picolinate and glucose tolerance formula (GTF) chromium, natural zinc picolinate, PABA, pantothenic acid, vitamin C from pure ascorbic acid, liver tablets, selenium, manganese, calcium-magnesium, primrose oil, B-complex, extra B-3 and B-6 and a natural multivitamin. Now, I supplement with a daily multiple vitamin, eight to ten grams of pure ascorbic acid for vitamin C, liver tablets, vitamins A and E and a B-complex. However, my supplementary needs are not universal; everyone has different nutritional needs so I always tell people to consult a doctor first and then a nutritionist to decide what their health needs are.

I also came to realise that the sugar-free, fat-free concept is a deadly deception. Fake foods contain chemicals that are attached to neuro-transmitters, such as isolated amino acids. These chemicals trick the body into believing it is eating or drinking real sugar or fat by transmitting false signals that the body has just received real food. But, what's inside the fake food is nothing. Man-made foods are "built" in laboratories. They are manufactured concoctions with little nutritional value. Just because they are free of fat and calories doesn't mean they are good for you. They are designed to fool the body.

A daily diet of nothing starves the body. A person cannot survive on a diet of fake food and chemical fillers. You may lose weight by tricking your body, but you'll eventually get sick. Even though you're eating, you're not supplying the nutrition your body demands to prevent disease.

Even today, standard medical tests completely ignore nutritional "malnutrition." Deprivation of nutrients will show up on standard laboratory tests only after a body organ has been damaged or a tumour has formed. There is no prevention practised. However, if you have malnutrition, some of the symptoms will be: headaches, PMS, seizures, dry skin and rashes, eye problems, mood disorders, depression, hypoglycaemia, intestinal problems, stomach-aches and fatigue. These can all be indications of potential nutritional deficiencies.

Sweet Poison

Because of the severity of my illness, my thyroid gland was damaged. My white blood cell count was dangerously high, which meant my immune system was stressed beyond helping itself. White blood cells aid the immune system. I discovered the best way to naturally stimulate my immune system was to saturate my blood with vitamin C. I have since learned a lot about this lifesaving vitamin.

Unlike most animals, the human body is unable to produce its own vitamin C. Humans, guinea pigs, apes and a species of bat found in India are the only animals known to be unable to produce their own vitamin C in their liver.

A 150 pound animal produces an average of fifteen grams of vitamin C every day. When the animal is stressed, it increases the amount of vitamin C as needed to meet the demands on its body. Some animals have been known to produce up to one hundred grams of vitamin C in a day when put under extreme stress. Humans depend exclusively on their diet for their supply of vitamin C.

Vitamin C is essential for good health. It is important for normal growth and development, for collagen production, connective tissue formation and healthy skin. It aids in the healing of wounds, in recovery from surgery and assists adrenal gland function and hormone production (especially in times of high stress). It also helps in proper cholesterol metabolism, proper iron absorption, bile production for good digestion, alcohol, drug, and smoking detoxification, and protects against pollution and free radicals.

Because of the poison saturating my body, I needed all the vitamin C I could get. As pollution from the environment and chemicals within our food increases, people need to supplement their diets with vitamin C and antioxidants more than ever before.

Toxins activate certain enzymes in the body. These enzymes transform the toxins into water soluble substances that can be excreted. Drinking lots of water and sweating during exercise helps eliminate these toxic solutions. However, if the toxic load gets high enough to saturate the fatty tissues, the toxins will begin to deposit in the body organs and in brain tissues. Over time, the contamination accumulates, creating a kind of human sludge. You can see this same type of sludge along river beds and shorelines, but you can't see it inside of your body. This doesn't mean it's not there! This sludge slows the oxygen flow and inhibits the ability of nutrients to reach individual cells. Due to the inability to breathe and be nourished, the cells mutate, making them susceptible to cancer and other diseases.

Nutrients within our bodies protect our cells from the damage toxins cause. They aid in eliminating toxic chemicals. If pollutants continuously enter your body, more and more nutrients are needed. The cleansing cycle goes on until the body is completely purified. And it can purify itself if it is supplied with enough of the right nutrients and no longer flooded with food poisons.

With the amount of synthetic chemicals in processed foods today, it is harder and harder to eat enough live enzymes, vitamins, and minerals to stay detoxified and nutritionally balanced. Pollution is increasing to such dangerous levels that all animals are absorbing damaging levels of toxins. It is overwhelming when you add it all up. We are bombarding our bodies with poisonous chemicals from inside and out. Frighteningly, pollution can accumulate to a point that it begins to burn, spontaneously erupting into a toxic fire scorching not only its source, but incinerating all the life around it. Contamination of the body not only damages us physically, it damages the mind, destroys a human being's self awareness, self-esteem, and pride, too. It burns the natural environments of self-worth, intentions, and decimates the fertile grounds of creativity.

When a river dies of pollution, its flow stops, as well as its life force. Plants, animals and people living by the riverbanks lose their life forces, too. Pollution causes disease in adults and deformities in children. No matter where we choose to live, if we live in a contaminated environment, our outlook on life may be polluted, too.

Chief Oren Lyons of the Onondaga American Indian tribe stated in the book 'Wisdomkeepers', "One of the natural laws is that you've got to keep things pure. Especially the water. Keeping the water pure is one of the first laws of life. If you destroy the water, you destroy life."

During the Global Forum on Environment and Development for Survival held in Moscow in January 1990, Onondaga Clan Mother, Audrey Shenandoah, delivered a keynote address in which she defined nature. She stated, "There is no word for 'nature' in my language. Nature, in English, seems to refer to that which is separate from human beings. It is a distinction we don't recognise. The closest words to the idea of 'nature' translate to refer to things which support life. It is foolish arrogance for humans to think themselves superior to all life-support systems. How can one be superior to that upon which one depends for life?

"I would urge the whole concept of nature be rethought," she continued. "Nature, the land, must not mean money; it must designate life. Nature is the storehouse of potential life of future generations and is sacred... Society needs to prioritise life-supporting systems and to question its commitment to materialism."

There is a strong analogy between what we are doing to the earth and to ourselves. Filtering factory emissions, recycling garbage, and saving baby seals are but meagre measures to reverse environmental damage already done. Human beings' abuse of nature has and is rapidly evolving. We can't afford to pollute the Earth as we are doing today. Nor can we afford to pollute our bodies without destroying ourselves. Car emissions, nuclear waste, landfills, cigarette smoke, silicone breast implants and aspartame are all poisons.

I know now that there are great similarities between a mountain stream bubbling with clear blue water and a river of rich red blood flowing within my veins. Life is life, and pollution is pollution inside or outside the body. And, in a polluted world, the corruption that ensues breeds disease. Human beings have inherited a responsibility to nurture a safe personal environment to secure the well-being of adults and most of all, the children of the future. We must not fail that commitment.

Bring your life into synch' with nature. Simplify it before a serious illness forces you to, as mine did, and you will be far ahead. Ever since I have followed that law of nature, my life has been more fulfilled.

PRIORITISE YOUR POISONS

Sweet Poison has undergone many changes. Finally, I was simply keeping a journal of a strange disease that accosted me, appearing out of nowhere. I wanted to understand the cause of my life-threatening case of Graves' Disease. I wanted to know why I was sick, but no one could provide any answers. Perhaps writing down my experiences would help me decide what to do about my strange illness, I told myself. Indeed, it did just that!

A good friend of mine, who is also a writer, suggested I journal my early Graves' symptoms, my experiences in the hospital, and the convoluted details of my recovery. "Even though they seem important now, you'll forget the small details over time," she said. "Write them all down because you'll be amazed at the little things you won't remember."

She was right. Looking back on it later, I realised I had forgotten how horrible I looked during my illness, how depressed I became and that I was suicidal. Later, I was grateful that I documented every detail from beginning to end.

At first my journal was a blueprint of my investigations. When I discovered the chemical sweetener aspartame caused my illness, my journal served as a personal reference for Sweet Poison. I thought sharing my encounter with aspartame could help others with similar experiences. As I became more involved in the aspartame issue, however, I discovered it was not that simple. Many people were ill but few even suspected the cause.

I wanted to speak publicly about aspartame and my personal experience with Graves' disease. Mary Stoddard suggested I submit magazine articles about my story. I decided to write a book instead. I dedicated over seven years to researching and writing 'Sweet Poison'. During those seven years I also became an active crusader. It was hard, but rewarding. It is daunting to go up against a big corporation. In some instances, it is even illegal. In Texas, where I live, the state legislature passed House Bill 722 into law in 1995. HB 722 states that anyone who speaks out against a perishable food product is guilty of a misdemeanour and subject to several thousand dollars in fines. The law gives corporations permission to sue individuals who make statements against their products. This is a serious matter and a potentially dangerous one for individual consumers.

The bill sat on Governor George Bush's desk for several weeks waiting to be signed into law. His office was inundated with phone calls from concerned consumers, health store owners, homeopathic doctors, nutritionists, wellness centres and the like, all protesting his signing this bill. Governor Bush was heavily lobbied by corporations to sign. His action: Governor Bush didn't do anything! He neither signed nor vetoed the bill. So, the bill automatically became law. This is the law that the recent beef industry lawsuit against Oprah Winfrey was based on. So, in order to bring my message to the public, I have had to study the laws of other states and even other countries. As well as becoming knowledgeable about the law, I have tried to learn as much as I could about other necessary fields.

I recently received an invitation from the American Academy of Environmental Medicine to attend a four-course programme designed to teach medical practitioners how to recognise and repair dysfunctional biological mechanisms and how to implement preventive, cause-oriented medical care.

Environmental medicine is an emerging field in medicine today, necessitated by situations just like mine. Aspartame is one of many unnatural chemicals the body doesn't need. Because it is not a natural food source, the body won't accept aspartame as food, no matter how hard you force it. Aspartame and other unnatural chemicals aren't used by the body – the body feeds exclusively on natural nutrients. If chemicals consumed in foods are not immediately excreted through sweat, faeces, or urination, they deposit inside the body as toxic waste. Doctors need to be environmental engineers when it comes to healing illnesses caused by these types of toxic loads.

As a certified nutritionist, I am afforded the opportunity to share what I've studied and learned, to help those around me. However, it is sometimes overwhelming to meet so many people dealing with problems caused by aspartame. Over eighty percent of my clients experience some form of health problem associated with aspartame: weight gain, hair loss, severe medical ailments, eating disorders, mental disorders, or fatigue. Some readily eliminate aspartame from their diets and their symptoms quickly disappear. But, many are addicted to aspartame and are unable to withdraw from diet products without suffering severe headaches, cramping, nausea, and diarrhoea. Sadly, they require a withdrawal programme similar to those for alcohol and drugs.

The nutritional work I do today is the direct result of all I experienced during my horrific reaction to aspartame. I learned that withdrawal from aspartame not only requires exceptional nutritional counselling, but significant emotional support as well.

I had little or no support when I first found out I had Graves' disease. Nowhere to turn for help; no advice, other than to destroy my thyroid gland. Before I met Mary Stoddard, there was nothing to confirm my suspicions about aspartame. There was no available information connecting aspartame with Graves' disease. Essentially, I stood up to my doctor alone. Unaware of my options, I desperately held onto a belief in myself and in my basic instincts. This degree of self support was hard to sustain. But I did it.

Cynthia Copeland Lewis wrote in her book, 'Really Important Stuff My Kids Have Taught Me,' "Build the base of your block tower wider than the top." So true! Everything needs a good foundation to be strong and long lasting. Good communication is an excellent foundation for supporting difficult situations, like a diagnosis of Graves' disease. On the contrary, a lack of communication results in limited support. There was, and still is, a definite lack of communication concerning the dangers of aspartame. Unfortunately, other areas of health and medicine suffer from this same problem.

My friend Meloni was recently diagnosed with cervical cancer at age thirty-two. Her doctor quickly performed two biopsies, informed her of the malignant outcome, and rushed her into surgery the next day. Before she had the chance to accept what was happening or was able to entertain any possible alternatives, her cervix and uterus were gone forever.

Snap decisions can sometimes save lives, but such changes may be, on many occasions, the wrong ones. Meloni needed time to consider her fate. Time to prepare for the surgery and her recovery. Time to gather the support she needed to help her through the worst trauma of her life. Time to grieve her loss. Time to gather emotions hastily left behind. Time and open communication to learn of all the available options before jumping into a decision. Unfortunately for Meloni, she wasn't given time. When I was hospitalised, I needed time to decide the fate of my Graves' disease. I'm glad I took that time, too. If I hadn't, I may never have discovered the actual cause of my disease: aspartame.

Now that my experience with aspartame is far enough behind me, I can thank my illness. I never want to go through it again, but my encounter with a deadly disease with "no known cure" proved to be the toughest, yet finest, teacher I have ever had. I learned you cannot unequivocally trust agencies like the FDA to protect you. I realised how far from the basics of living I had strayed. How unhealthy my lifestyle was. It showed me how processed foods have become, particularly sugar-free chemical sweeteners. It became obvious to me that being thin and good-looking is more important in our society than having good health. I learned how to face death.

I have made my misfortune my good fortune. I have become more philosophical and at the same time, more questioning. With every new experience, I have tried to grow. The more I learn, I tell myself, the stronger I become. And I have tried to become less of a perfectionist and more forgiving. I repeat to myself that I am only human and doing the best I can. I'm learning as I go. I must remember there are some things which can't be fixed, to quietly let go and to trust.

Embracing this philosophy, I decided that it was finally time to dissolve my unhappy marriage. Before my illness, my husband and I were growing apart. I kept myself so busy that I avoided the reality in front of me. When I got sick, I felt sad that he was totally disinterested, but my marriage wasn't important at the time. I was fighting to survive and trying to understand why I got sick, how to get well and how to cope with the injustice of what could have happened to me. My life would never be the same. I was not the same.

I devoted so much of my soul to my recovery and to Sweet Poison, I neglected my marital problems. Now it was too late; my husband and I had drifted too far away to find one another again.

One day while standing in the kitchen, I simply said to Chuck, "I want a divorce." With no argument or comment, Chuck simply responded, "Okay."

And that was that. We talked to the kids about why we were getting divorced, divided our things, and he moved out. After eighteen years of marriage, in less than ninety days I returned to my christened identity, Janet Starr Hull. I felt relieved that I had finally faced the truth and oddly exhilarated.

It's really sad how things work out sometimes. I feel as if I've never been married. But, I do have three dynamic and wonderful sons I'm most grateful for.

I learned there is so much more that is necessary to heal a physical illness, or a failed marriage for that matter, than a quick fix. Healing was a quiet process and a gentle leap of faith. I turned a negative experience into a positive one. It's all in how you look at it. The more I healed, the more I wanted to help those around me.

I have a good friend, Gail Shaddock, who is a dynamic psychotherapist in Dallas. Gail and I work together periodically. One afternoon, Gail called from her office asking if I could see one of her clients as soon as possible.

Carole Carter was already seeing a psychiatrist and a medical doctor when she began counselling with Gail. She was suffering multiple seizures every day. In her thirties, on disability, and no longer able to work, Carole moved back home so her mother could take care of her. Her doctors ran all the appropriate tests, but found nothing wrong with her.

They recommended she get counselling, concluding she was causing the seizures herself. "Emotional seizures" was the term they used.

"Is there such a thing?" asked Gail.

"I don't think so," I replied. "I've never heard of anyone mentally causing their own seizures."

One day, Carole walked into Gail's office holding a diet drink. When Gail saw her, she cried out in surprise, "What do you have in your hand?"

Taken aback, Carole meekly answered, "A diet soda. Why?"

"Don't you know diet drinks can be bad for you?" Gail replied, her energy piqued. "You need to talk to Jan."

Gail immediately arranged a meeting. Gail introduced me to Carole, explaining who I was and what I was doing at her therapy session. Carole looked petrified. Gail felt the source of her "emotional seizures" could be aspartame. Carole had no idea what was happening.

Her history was one I hear too often. Carole drank in excess of twelve diet colas a day. She said she had a diet cola in her hand all day long when her seizures began. They could not identify the cause of her seizures. Doctors prescribed a variety of medications, which provided temporary relief, however none were permanent solutions. As with my Graves' disease, the symptoms were easily masked by medication. Until the cause was identified, there was little chance of permanent recovery.

Carole was willing to quit all aspartame, replacing her diet drinks with water and juice. She agreed to improve her general diet and to begin a natural supplement program. I suggested she exercise regularly. Simply walking was fine, just as long as she did something regularly. Gail and I watched her closely for the next several weeks. I informed Carole's medical psychiatrist of my dietary recommendations.

"I believe the aspartame is the source of Carole Carter's seizures. I feel very strongly about it," I conveyed to her psychiatrist. Much to my surprise, the psychiatrist agreed with me. We three watched Carole's progress, and waited.

Immediately after removing all aspartame from her diet, Carole's seizures reversed themselves. Within seventy-two hours, her multiple seizures stopped completely. After three weeks, she had as little as one seizure per week. After five weeks, she had so few seizures there was no pattern for my records. She was on her way to living a normal life again, seizure-free. There would no longer be thousands of dollars spent on disability and doctors visits. It was evident that she did not suffer from "emotional seizures." It was the aspartame all along.

Unfortunately, most people know little about aspartame. Found in over 5,000 foods, beverages, and medications, aspartame is in too plentiful a supply, and people are exposed to a lot of it without knowing of its dangers.

When I talk to people about aspartame, I usually begin telling them about the available research that shows aspartame's negative test results – tests no one has ever heard about. Then, I tell them about pilots passing out in their cockpits while consuming drinks and snacks that contain aspartame in flight. Most people are horrified at this. I also relate the results of the laboratory experiments that produced holes in the brains of mice. I tell them that, in January 1997, Dr. John Olney stated that there were more brain tumours recorded than ever before in human history, and that the rise had occurred over the past sixteen years. Aspartame came on the market in 1981, exactly sixteen years before 1997. Scientists like Dr. Olney who warned the FDA and the United States Senate in the 1970s and 1980s about seizures and brain dysfunction, foetal deformities, and tumours caused by aspartame were right all along!

I speak of cleft pallet deformations in babies. I always tell them about Patty Crain's death and how she'd put six or seven packets of diet sweetener in one glass of iced tea. I also relate the fact that the FDA has 10,000 complaints against aspartame in their files. I never forget to tell them about my Graves' disease. I explain what Graves' disease is and how my Graves' simply "went away" after I stopped using all products containing aspartame. Aside from these startling facts about aspartame, I also always warn them about how easily aspartame is hidden in our food, and discuss the importance of identifying and avoiding aspartame in the foods they buy.

I explain that when aspartame was marketed exclusively as NutraSweet/Equal, it was distinguishable by its prominent candy-like swirl on product labels. However, since the NutraSweet Company's patent expired at the end of 1992, aspartame can be manufactured by anyone and put into products without a conspicuous identifying mark. Therefore, even though all ingredients must be listed on food labels, aspartame is now much less noticeable in food products and can easily slip past your attention. I've had people say that when they buy diet or sugar-free products, they will be on the lookout for additives like aspartame.

People are usually shocked when I tell them that aspartame is being put into products that are not even labelled sugar-free, particularly in things like gum, mints, and yoghurts. I also tell them to beware of vitamins, medicines and other pharmaceuticals since they too, often contain aspartame. I tell them that often, sugar or corn syrup will be the first listed ingredient on a product's label, but somewhere down the list, aspartame may appear. One tip I give to people to help them quickly spot hidden aspartame is to look for the warning, "Caution: Phenylketonurics." Since phenylketonurics cannot process phenylalanine, which comprises half of aspartame, those afflicted can suffer permanent brain damage when they ingest phenylalanine.

I usually try to turn the discussion to general nutrition and good health. We talk about how to replace "diet" drinks in the diet with water, natural foods, and lots of exercise.

The conversation often turns to their health problems. The complaints I hear are multifold. Headaches, mostly. Migraines. PMS. Mood swings and depression. Seizures. Dry skin. Eye problems. Fatigue. And on and on. When the topic shifts to children, I tell them of the tragic story of little Katrina Carradine whose needless suffering should never be endured by any child.

For myself and my family, I will never accept any more fake chemicals in our food, only fresh and natural items. I also drink plenty of water and get daily exercise and practice lots of fire fighting. Once I naively assumed everyone knew they should do these things. I was wrong. I have learned that many people have much more difficulty giving up diet foods than I realised.

Since aspartame came on the market, my neighbour has had a diet drink in her hand every day. It was nothing for her to drink at least a dozen diet colas at work alone. Having met me, she stopped using aspartame – or tried to, anyway. She had no idea she would have bad withdrawal symptoms. She stopped aspartame cold turkey, and became violently ill. She suffered severe abdominal cramping and horrendous headaches. Drinking a diet cola relieved those symptoms, but induced worse ones. What she experienced was not unusual while trying to withdraw from aspartame.

While withdrawal from aspartame is difficult, it is necessary and should be combined with healthier eating by adopting a natural, whole foods diet. The basic steps back to good nutrition are initially the same:

• Read food labels and avoid all aspartame. Aspartame is in over 5,000 products. That's a lot of labels to read. Presently, aspartame is found in countless products without the familiar NutraSweet swirl. Since the NutraSweet Company's patent expired, generic aspartame can be found anywhere.

• Drink water. Water is the most important life-giving substance in the body, and the body desperately depends on it. People who have a diet drink in their hand all day at work are drinking far too many colas and not enough water. Carry bottled water around instead. On the average, a human eliminates eight cups of water a day. Put back into your body what you give up. Iced tea, diet drinks, coffee or powdered punch don't count as water sources. Pure water is what the body must have to properly flush toxins from the system.

• Get a hair analysis. Vitamin and mineral levels inside your body register in the hair. Toxins also show up in the hair. I recommend everyone have at least one hair analysis in their lifetime. A blood test identifies your blood type, a hair analysis shows what's inside the body. It reveals what vitamins and minerals you need, and possibly what toxins have accumulated within the tissues. Toxins stimulate allergies and can cause disease.

• Prioritise your poisons! Make a list of what you generally eat and drink. First eliminate what will kill you the fastest. If you smoke a pack of cigarettes a day plus drink coffee all morning, I recommend you quit smoking before giving up the coffee. Everywhere we turn, someone is telling us "This is bad for you," "That is going to kill you," "This could cause cancer." Now, I'm saying, "Aspartame may harm you, too." Sorry. Try not eating any products containing aspartame for thirty days. You may feel better, have fewer headaches, fewer mood swings. Maybe your appetite will return to normal, and you'll fill up more quickly after each meal. Like peeling the skin off an onion, keep peeling the chemicals away, layer by layer, until you have purified your diet. Eat like your grandparents – real food from the barnyard, vegetables pulled out of the ground, fruits, nuts and berries picked from the trees.
• Eat raw. If there's only time to grab quick food, keep lots of fresh vegetables, natural dry cheeses, whole grain crackers, pickles, natural yoghurts, tofu, fruits and nuts, water and fresh juice on hand. You may have to shop more often because the "real stuff" spoils more quickly, but that's because it's loaded with enzymes and live nutrients that do something for your health. Raw, steamed, natural and fresh. A nutritious quick meal lets you eat all you want. It's hard to gain weight eating only natural foods.
• Think of your food plate as a 'pie'. Seventy-five percent of that pie should be raw or steamed every meal. These 'fundamental' foods provide all the digestive enzymes the body needs, as well as the vitamins and minerals required to trigger the digestive chain reaction.
• Lay out a variety of fresh snacks after school or work. My kids love them, and they actually eat them! I snack on raw foods every day: apples, oranges, raw peanuts in the shell. Eat big meals early in the day. At night, your body processes what's left over. Try not to eat heavy food past five or six o'clock in the evening. For an evening snack, pop popcorn in cold-pressed seed oil with a little butter and sea-salt or drink a fruit smoothie.
• Maintain good digestion. Manufactured foods inhibit proper digestion. Supplement with a digestive enzyme rich in betaine hydrochloride: try a papaya, raw lemon, or a rich red wine. Proper digestion is critical to good health. Digestion is where "eating right and good health" begins. Remember, if you have stomach problems, you can support the stomach lining by eating raw cabbage at least three times a week to keep the stomach lining thick and healthy.
• Eat natural foods as opposed to manufactured, chemical foods.

Sweet Poison

Natural foods supply fuel needed to stay healthy, to keep your body moving and to stay mentally alert. Food converts to raw energy. All food passes through the same set of reactions, whether it's fast food or a raw carrot. What the food provides is the issue. Is your food rich in natural nutrients or is it full of chemical fillers and toxic by-products designed to "fake out" the body so you won't gain weight? Natural or fake, you decide.

Exercise! Move your body every day. It doesn't matter what you do, just do something. I used to teach aerobics twice a day. Now, all I have time to do is bike or walk two to three miles a day, plus my stomach crunches. What really counts: I do something every day, even if for ten minutes. Your body will depend on the consistency, and that's good. Exercise is a good way to replace bad habits, too. If you want to quit smoking, let's say, replace the habit with exercise. Find what type of exercise you will adopt permanently into your life, and never quit. Biking around the neighbourhood after work, walking around the block before the day begins, swimming ten laps every day, rowing while you watch the evening news, whatever – your body will come to depend on it.

• Avoid fake foods. Vitamin and mineral-rich foods are what I refer to as natural foods. They are mandatory to maintain good health. Fat-free, sugar-free doesn't provide 'natural'. When your food doesn't offer natural vitamins, your body has to supply them. When these supplies are depleted, you must replenish them. Not in fast food lines, not with diet colas, never with fake food substitutes, but with natural foods. Only with vitamin and mineral-rich foods can you sustain a good supply of the 'parts that keep the machinery going.'

• Use saccharin before aspartame if you must use an artificial sweetener. Many doctors, dentists, dieticians, and consumers support the relative safety of saccharin. Ultimately, I recommend using stevia or unprocessed sugar over saccharin. Remember, use natural products instead of artificial ones.

In 1991, approximately 14,555 food additives were on file with the FDA. Chemicals to enhance flavour, colour, preservation, and 'enrichment'. Aspartame was merely one of many chemicals bombarding our lives.

In 1995, other sugar substitutes, such as sorbitol, maltitol, lactitol, xylitol, acesulfame-K, and stevia, became more widespread. Of these, stevia is the only one that is natural. Aspartame, no longer the exclusive sweetener in manufactured food products, was now found in combination with various artificial sweeteners.

On June 4, 1997, the NutraSweet Company, now the NutraSweet Kelco Company, announced it's forthcoming 'NutraSweet 2000', a new chemical sweetener currently undergoing studies for FDA approval. The company confidently wrote that NutraSweet 2000 would be approved for public use, and they have good reason to believe it. After all, the FDA approved aspartame in 1981 despite evidence revealing it's dangers. According to the NutraSweet Kelco Company, NutraSweet 2000 "appears to be an improvement over original NutraSweet and is extremely potent." NutraSweet 2000 will be forty times sweeter than aspartame and seven thousand times as sweet as sucrose.

What can you do about aspartame?

• Set an example by changing your diet.
• Tell everyone you know.
• Talk to schools and day care centres.
• Offer to speak at parent-teacher meetings.
• Contact your local, state, and federal government representatives.
• If you see someone with a diet drink, ask if they have had any of the typical aspartame symptoms.
• Spread the word at your work.
• Distribute the ACSN and Pilot's Hotline number: 001 214-352-4268.
• Tell your doctor about the scientific research available.
• Register a complaint with the FDA, the FAA, the NutraSweet Kelco Company: No NutraSweet 2000!
• Return all food products with aspartame, opened or unopened, to your grocer. Tell him or her the products make you sick. The grocer can return them to the manufacturer for a store refund. The manufacturer should get the message. So will the grocer.
•   Spread the word on computer networks.
• Publish articles in newsletters at your church, place of work, or neighbourhood association.
• Set a personal example for health and wellness.

Become aware of all unnecessary chemicals hidden in your foods. Be a food detective. Purify your diet as much as possible and try to enjoy natural eating. Drink more purified water than any other beverage. Cut back on artificial food substitutes. Get back to the basics of eating and exercise regularly. And, WATCH OUT FOR THE SWEET POISON!

AFTERWORD

The world around us is a fragile place. It is home to many thousands of different types of plants and animals. These vary greatly from region to region, but all living things share a common dependency on the natural balance and health of our environment. It is precisely this natural balance between species and their habitats that is now being threatened by human kind's abuse and misuse of the Earth's precious resources.
– 'The Atlas of Endangered Species'

We must turn back to nature and learn from her how to live through her bounty and not consume artificial products. In his book, 'The Seasons of Life,' businessman and author Jim Rohn illustrated one of the best literary similes between man and nature. Rohn stated that people experience internal seasonal changes in the same way they observe external seasons. If people would learn to recognise and accept these natural changes, it would be possible to live a more rewarding life. Rohn wrote:

"The winter months are a time of rest and a mental preparation period for the work awaiting spring… This is the season to reflect on the past months, assessing what went well and what didn't. When spring arrives, your mind and body are rested and open with new ideas and the energy to carry those ideas through; to plant the seeds of creativity. By summer, expect to stay in the productive mode for the next few months, waking early and retiring late, as the time to activate your mind and creativity are now Even though the body works heard in the summer, much work remains to be done in the autumn for the harvesting from spring and summer and the preparation for winter's rest.

And, the cycle begins and ends and begins again for all living things. During each season since I've gotten well, I have taken the time to celebrate how far I've come since my illness and how much I have learned along the way. When I first got ill, my children were only three years, one and one-half years, and a few months old. As the dream of Sweet Poison materialises, they've turned fifteen, thirteen, and eleven. For most of their lives, they've known me as the "mom who fights aspartame." This book is as much theirs as it is mine. But even more, it is yours. For I have written this book to tell those of you I do not know that you must be vigilant to protect your health and the health of those you love so that neither you nor they will be seduced by those around the world procuring sweet poison.

APPENDIX 1

ORGANISATIONS AND RESOURCES

Aircraft Owners and Pilots Association (AOPA) 421 Aviation Way Frederick, Maryland 21701 Tel: (001 301) 695-2000. Dr. Stanley Mohler's book 'Medication and Flying: A Pilot's Guide' was published by and is available from AOPA.
The American Academy of Environmental Medicine (AAEM) Foresight America for Preconception Care 10 East Randolph New Hope, Pennsylvania 18938. Tel: (001 215) 862-4544 Fax: (001 215) 862-4583 The AAEM provides alternative health care treatments and therapies in order to provide affordable and effective health care to the public.
American Health Science University and National Institute of Nutritional Education (NINE) 1010 South Joliet, Suite 107 Aurora, Colorado 80012. Tel: (001 303) 340-2054 Tel: (800) 530-8079 Fax: (001 303) 367-2577 e-mail: nuted@aol.com – website: http://www.nines.com NINE is the oldest and most established school in the country offering programs in nutrition and updated educational information on current health issues.
Aspartame Consumer Safety Network (ACSN) World-wide Pilot's Hotline PO Box 780634 Dallas, Texas 75378. E-mail: Marystod@airmail.net – website: http:/ /web2.airmail.net/marystod Tel: (800) 969-6050 Tel: (001 214) 352-4268 Founded by Mary Nash Stoddard in 1987, ACSN offers books, videos, audio cassettes and general information about aspartame for media and the general public. The Pilot's Hotline is (214) 352-4268.
Body Ecology Diet Contact: Donna Gates 1266 West Paces Ferry Road, Suite 505 Atlanta, Georgia 30327. Tel: (800) 4-STEVIA Tel: (001 404) 352-8048 Donna Gates is a stevia representative and the author of Body Ecology Diet
Centre for Science in the Public Interest (CSPI) 1875 Connecticut Avenue, N.W, Suite 300 Washington, DC 20009. Tel: (001 202) 332-9110 Fax: (001 202) 265-4954, e-mail: cspi@cspinet.org – Website: http://www.cspinet.org This non-profit organisation is supported by over a million members through its newsletter, 'Nutrition Action'. CSPI considers aspartame to be the third worst food additive on the market today.
Children Harmed By Aspartame, PO Box 971 McKinney, Texas 75050-0971. E-mail: jshull@sweetpoison.com – website: http://www.sweetpoison.com If you or someone you know has a child who has been permanently damaged by the artificial sweetener aspartame, please contact Sweet Poison's website, www.sweetpoison.com, for more information on how to receive medical and financial assistance. Or please write to Children Harmed By Aspartame.
Cognitive Enhancement Research Institute (CERI), PO Box 4029, Menlo Park, California 94026. Tel: (001 650) 321-CERI Fax: (001 650) 323-3864 Website: http://wwwceri.com CERI – nucleus of information concerning contemporary health issues. Their newsletter, Smart Drug News, is available by contacting

CERI or visiting their web site.

Department of Health and Human Services US Food and Drug Administration Washington, DC 20204. General Information: (800) 532-4440 Website: http://www.fda.gov To report problems including adverse reactions to any food product excluding meat and poultry call the FDA Food and Seafood Information Line (800) FDA-4010. To report problems to a consumer complaint co-ordinator in your area, contact your regional FDA district office.

Emergency Response Guidebook, National Institute for Occupational Safety and Health Pocket Guide to Chemical Hazards. Both pocket guide books are good references to use when researching the various chemicals and additives in the foods that you eat as well as the chemicals that are around your home and in the environment. Both are available by calling (800) 327-6868.

James B. Hays, MD, General Practice, Surgery, FAA Designated Medical Examiner, 101 C. South Park Drive, Brownwood, Texas 76801, Dr. Hays is an FAA Designated Medical Examiner who works with pilots who have had reactions to aspartame while flying.

Erik Millstone, Ph.D., Science Policy Research Unit, Mantell Building, Sussex University, Brighton BN1 9RF, England, Tel: (44) (0)1273-877380 Fax: (44) (0)1273-685865 e-mail: E.P.Millstone@sussex.ac.uk – website: http://www.susx.ac.uk/spru/

The NutraSweet Company, Box 900, 1751 Lake Cook Road, Deerfield, Illinois 60015 Tel: (001 312) 940-9800 Tel: (800) 321-7254 Website: http://www.nutrasweet.com

NutriVoice, PO Box 946, Oak Park, Illinois 60303, Editor: Barbara Mullarkey, NutriVoice is a health newsletter edited by Barbara Mullarkey.

Palm Beach Institute for Medical Research, Inc. H.J. Roberts, MD, P.A. PO Box 17799 West Palm Beach, Florida 33416 Tel: (561) 432-4774

Starr Resources International (SRI), PO Box 971, McKinney, Texas 75070-0971, E-mail: jshull@sweetpoison.com – website: http://www.sweetpoison.com Founded by Janet Starr Hull, SRI offers nutritional counselling, personal hair analysis, motivational seminars and has designed a programme for aspartame withdrawal.

Stevita Company, Inc. 7560 Highway 287, Suite 100, Arlington, Texas 76017 Tel: (001 888) 783-8482 Stevita is a distributor of stevia products.

Swankin & Turner, Attorneys at Law, 1424 16th Street, Suite 105, Washington, DC 20036 Tel: (202) 462-8800 Fax: (202) 265-6564 James S. Turner represents individual consumers, consumer groups and environmental groups concerning matters in food, drug and environmental issues.

APPENDIX 2

INTERNET RESOURCES

http://www.aspartame.com

http://www.ceri.com – Cognitive Enhancement Research Institute

http://www.cspi.net.org – Centre for Science in the Public Interest; offers information and reports on nutrition and health.

http://wwwdhhs.gov, Department of Health and Human Services

http://www.dorway.com/stevia.html – Provides information on stevia and other natural sweeteners, including where stevia can be purchased.

http://www.fda.gov – The Food and Drug Administration

http://www.holisticmed.com/aspartame/

http://www.lbmic.com/smith. html – Provides information on Dr. Lendon Smith.

http://www.nines.com – National Institute of Nutritional Education

http://www.nutrasweet.com – The NutraSweet Company

http://www.sweetpoison.com – Starr Resources International & Children Harmed By Aspartame

http://www.susx.ac.uk/spru/ – Dr. Erik Millstone at Sussex University.

http://www.trufax.org/research/aspart.html – Provides information on aspartame.

http://www.trufax.org/research/ f26.html – A listing of pharmaceutical company products that contain aspartame as compiled by Leading Edge Research.

http: //www.vitawise.com/aspfacts.htm – Written by former NBC physician Dr. Lendon Smith.

http://web2.airmail.net/marystod/ – Aspartame Consumer Safety Network (Mary Nash Stoddard)

APPENDIX 3

Testimony Presented to US Congress Regarding Aspartame

PALM BEACH INSTITUTE FOR MEDICAL RESEARCH, INC., A NON-PROFIT
CORPORATION WEST PALM BEACH, FLORIDA, H.J. ROBERTS, MD

THE ASPARTAME PROBLEM, STATEMENT FOR SENATE HEARING
NOVEMBER 3, 1987

I am concerned over the escalating number of aspartame-related complaints
documented by or reported to interested investigators, the FDA, and
consumer groups such as Aspartame Victims And Their Friends and the
Community Nutrition Institute.

The most serious reactions I have encountered among 551 aspartame reactors
in my nation-wide computerised study include severe headaches (45.2%),
dizziness (39.4%), decreased vision or blindness (27.9%), epileptic attacks
(17.8%), profound confusion and memory loss (28.5%), extreme depression
(25.2%), and marked personality changes (16%). Hives, other rashes, itching,
mouth reactions, hyperactivity, extreme fatigue, insomnia, menstrual
disorders, diarrhoea, intense thirst, ringing in the ears, aggravated
hypoglycaemia, loss of diabetic control, severe weight loss (suggestive of
anorexia nervosa), a paradoxical gain of weight, and joint pains also have
been prominent. These observations, which have been reported in the
references listed, are supported by FDA data.

An unexpected finding has been the occurrence of reactions to aspartame in
the relatives (up to seven) of twenty-two percent of persons so affected.

The causative or contributory role of aspartame is supported by (1) the
gratifying relief or disappearance of these complaints (except for blindness
and severe neuro-psychiatric features) after its cessation; and (2) their
predictable recurrence on re-challenge.

Damage to the retina, optic nerves and brain can be caused by methyl alcohol
in heavy consumers. (Aspartame promptly releases ten percent methanol by
weight when metabolised.) Other complications may reflect the combination
of high levels of phenylalanine within the brain, altered neurotransmitter
function, severe caloric reduction and increased release of insulin and other
hormones. Recent data also raises the possibility that selective amino acid
imbalance of phenylalanine and aspartic acid may accelerate or aggravate
Alzheimer's disease, Parkinsonism, behavioural abnormalities and other
neuro-psychiatric afflictions.

The problems of aspartame reactions assume even greater importance for certain high-risk groups. They include the many children, pregnant women, older persons and patients with diabetes, migraine, hypoglycaemia, epilepsy, allergies and liver or kidney disease who consume large amounts of such products.

Severe deficiencies pertaining to the evaluation, licensing, promotion and surveillance of aspartame must be addressed in the face of this potential "recipe for disaster." They encompass (1) the need for accurate clinical data collected by the responsible governmental agencies, including an analysis of tumours (especially brain and urinary bladder) and blood disorders, (2) the arbitrarily high maximum allowable daily intake, (3) the prior resistance by Congress to enactment of legislation that would mandate the proper labelling of aspartame products, (4) the failure to challenge the alleged safety and effectiveness of aspartame in massive promotional campaigns, especially for long-term weight loss, and (5) unfounded bias against saccharin based on limited and controversial studies concerning urinary bladder tumours in male rats.

It is now the burden of regulatory agencies that have minimised or denied the accumulated clinical data to convincingly explain them in terms other than isolated idiosyncratic reactions.

The informational gap caused by the biases and other misgivings of reluctant editors, publishers and chairmen of major medical conferences is a public health and scientific disservice.

Concerning my credentials, I have been certified and re-certified by the American Board of Internal Medicine, and am a member of the Senior Active Staff of the Good Samaritan Hospital and St. Mary's Hospital in West Palm Beach. I have authored more than 200 original scientific publications and five texts, and am listed in 'The Best Doctors in the US' and 'Who's Who in the World.'

REFERENCES

1. Roberts, H.J.: Is aspartame (NutraSweet) safe? On Call(publication of the Palm Beach County Medical Society). January 1987, pp. 16-20.
2. Roberts, H.J.: Neurologic, psychiatric and behavioural reactions to aspartame in 505 aspartame reactors. In Wurtman, R.J., Ritter-Walker, E. (eds) Proceedings of the First International Conference on Dietary Phenylalanine and Brain Function. Washington, DC, 1987, May 8-10, pp. 477-481.
(Signed) H.J. Roberts, MD

EMORY UNIVERSITY SCHOOL OF MEDICINE, Department OF Paediatrics, DIVISION OF MEDICAL GENETICS, ATLANTA, GEORGIA. LOUIS J. ELSAS, II, MD

STATEMENT FOR THE LABOR AND HUMAN RESOURCES COMMITTEE, US SENATE

I have considerable concern for the increased dissemination and consumption of the sweetener, aspartame (1 methyl N-L-a-aspartyl-L-phenylalanine) in our world food supply. This artificial) dipeptide is hydrolysed by the intestinal tract to produce L-phenylalanine which in excess is a known neuro-toxin.

Sweet Poison

Normal humans do not metabolise phenylalanine as efficiently as do lower species such as rodents and thus most of the previous studies on aspartame effects on rats are irrelevant to the question, "Does phenylalanine excess occur with aspartame ingestion?" and if so, "Will it adversely affect human brain function?"

THE ROBERT W. WOODRUFF HEALTH SCIENCES CENTRE

Preliminary studies in my laboratory provide tentative positive answers to both questions. Many studies of both acute and chronic ingestion of 34 mg aspartame/kg/day have demonstrated a two to five fold increase in semi-fasting blood phenylalanine concentrations (from approximately 50 to 250 M) without concomitant increases in tyrosine or other amino acids. The degree of increase by normal humans depends on several variables including the efficiency of gut transport, liver utilisation and growth rates. It was thought by many scientists and clinicians that this degree of blood phenylalanine increase would not affect brain function. However, currently available information indicates that this is not true.

1)   In the developing foetus such a rise in maternal blood phenylalanine could be magnified four to six fold by the concentrative efforts of the placenta and foetal blood brain barrier. Thus, a maternal phenylalanine of 150 M could reach 900 M in the developing foetal brain cell and this concentration kills such cells in tissue culture. The effect of such an increased foetal brain concentration in viro would probably be much more subtle and expressed as mental retardation, microcephaly, or potential certain birth defects. In the rapidly growing post-natal brain (children of 0-12 months) irreversible brain damage could occur by the same mechanism. In the adult, we have found that changes in blood phenylalanine in these concentration ranges are associated with slowing of the electroencephalogram, and prolongation of cognitive function tests. Fortunately, these effects on the mature brain are reversible but provide clear evidence for a negative effect on sensitive parameters of brain function.

In view of these new (and confirmation of old) research findings I suggest the following:

1)   Immediate quantitative labelling of all aspartame-containing foods, so the consumer will know how much phenylalanine he/she is ingesting.

2)   Declare an immediate moratorium on addition of aspartame to more foods and remove it from all low protein beverages, foods and children's medications.

3)   Provide funds not controlled by industry to:

a)   Allow active surveillance for potential side-effects of aspartame on new-borns whose mothers dieted with aspartame-containing foods.

b)   Allow active evaluation of other users whose complaints cannot be adequately studied at present. Clarify the dose relationship and mechanisms by which L-phenylalanine affects human brain function.

Respectfully submitted, (signed) Louis J. Elsas, II, MD Director, Division of Medical Genetics Professor of Paediatrics

UNIVERSITY OF FLORIDA COLLEGE OF PHARMACY, DEPARTMENT OF
PHARMODYNAMICS, J. HILLIS MILLER HEALTH CENTRE, RALPH DAWSON, JR.,
PH.D. GAINESVILLE, FLORIDA

STATEMENT TO SENATOR EDWARD M. KENNEDY, CHAIRMAN, SENATE
COMMITTEE ON LABOR AND HUMAN RELATIONS

I am writing as a concerned scientist in regards to the forthcoming hearings
on the safety of aspartame. I have published over 10 scientific papers in
refereed journals on the neuro-toxic effects of monosodium glutamate.
Glutamate and aspartate share the ability to activate excitatory amino acid
receptors on nerve cells and function normally as neurotransmitters, or at
excessive extracellular concentrations can cause the death of nerve cells. I
wish to address several concerns that I feel are vital to a comprehensive and
scientifically sound assessment of the safety of aspartame as an artificial
sweetener (food additive).

The specific concerns are as follows:

1)   Excitatory amino acids act synergistically to cause both
pharmacological and neuro-toxic effects. Thus, to consider only aspartame
intake in the calculation of risk to neuro-toxic damage is completely
erroneous. Therefore, daily intake of natural foods high in aspartic and
glutamic acid must be considered in addition to foods containing
monosodium glutamate (MSG) as an additive. A second point in this regard is
the ubiquitous presence of aspartame in the present food supply (children's
cereal, soft drinks, snack bars, frozen popsicles, etc.). An accurate assessment
of total possible intake of aspartame plus other dietary sources of glutamic
and aspartic must be made.

2)   The correct identification of individuals in special risk groups must
be made. Possible candidates at special risk to aspartame include: neonates,
the elderly, diabetics, individuals suffering from Chinese Restaurant Syndrome
and those individuals at risk for cerebral ischaemia, epilepsy or neuro-
degenerative disorders (olivopontocerebellar atrophy, Alzheimer's Disease,
Huntingdon's Disease, etc.). Glutamate and aspartate play a pivotal role as
neuro-transmitters, and metabolic disorders involving glutamate or aspartate
may be involved in a variety of neurological disorders. Thus, the unregulated
dietary intake of excessive amounts of these excitotoxins may exacerbate pre-
existing neurological conditions. An extensive effort should be made to
identify any groups of individuals for which aspartame ingestion would have
deleterious consequences. There may also be individuals that show metabolic
intolerances to aspartame. This intolerance could be to the aspartic acid as
already discussed, or to the phenylalanine (such as PKU patients) or
methanol. The possibility that aspartic acid and methanol could act
synergistically to produce visual disturbances must also be considered
(aspartic acid can cause retinal damage and methanol damages the optic
nerve). Individuals with defects in key enzymes that inactivate these
metabolises of aspartame would be at special risks to various toxic sequelae.

3)   A most disturbing action of aspartic acid, given the presence of
aspartame in so many products consumed by children, is its ability to

stimulate hormone release from the anterior pituitary gland. The aspartic acid analogue, N-methyl-DL-aspartate (NMDLA) has been shown to induce large and rapid elevations in plasma levels of luteinizing hormones, follicle stimulating hormone and prolactin. Growth hormone release is also altered by glutamate administration of NMDLA infusion. These actions are not toxic in the classical sense but the regulated pulsatile release of anterior pituitary hormones is essential to the normal growth and sexual maturation of children, as well as the control of the reproductive cycle in mature females. These neuro-endocrine effects of a food additive must be considered in regard to the safety of the product. It cannot be emphasised enough that aspartic acid is a neurotransmitter that functions importantly in many completely integrated physiological processes. Phenylalanine can be converted to tyrosine, which serves as a precursor to the catecholamine (dopamine, norepinephrine and epinephrine) neurotransmitters. Thus, even at doses of aspartame that are non-toxic there can be significant pharmacological actions such as disruption of hormone release and changes in brain function.
Specific Recommendations:

1.   Detailed and comprehensive statistical evaluations of consumer complaints concerning aspartame and disclosure of these findings to the scientific community.

2.   Revamped safety testing to include both aged and neonatal rodent and primate testing. The evaluation of both toxic and pharmacological actions of aspartame at levels that approximate total excitotoxin (glutamate and aspartate) intake.

3.   An aggressive research program to identify any special groups at risk to aspartame including: a. the elderly. b. children. c. diabetics. d. victims of the Chinese-Restaurant-Syndrome. e. epileptics. f. individuals with cerebral ischaemia or hypoxia. g. individuals at risk for neuro-degenerative disorders.

4.   Identification of the incidence of metabolic intolerance to aspartame in the general population. How many people show adverse reactions to aspartame and what is the metabolic basis?

5.   Federal funding of research on the neuro-pharmacological and neuro-endocrine actions of long-term aspartame ingestion. Do we have a potential long-term public health concern with unknown actions of chronic daily aspartame ingestion?

6.   Public education (warning labels?) concerning the issue of aspartame safety. There is not uniform agreement among the scientific community regarding the safety of either aspartame, saccharin or cyclamates. The public must not assume FDA approval (under duress due to public pressure at the time of approval) assures that a product is biologically innocuous. Future public hearings would certainly be in order.

My concerns are those of a scientist that feels not enough data are available. I do not wish to see consumers as subjects in a large scale clinical trial of a product. I am unable at present to say what the magnitude or scope of any adverse consequences will be to the long-term introduction of aspartame (in combination with MSG) into the food supply. I only hope we have the opportunity to aggressively seek the answers to the issues I have raised using scientific criteria unencumbered by economic, political or bureaucratic

artifice. Sincerely,
(Signed) Ralph Dawson, Jr., Ph.D. Assistant Professor

UNIVERSITY OF CALIFORNIA, LOS ANGELES, UCLA SCHOOL OF MEDICINE,
LOS ANGELES, CALIFORNIA, WILLIAM M. PARDRIDGE, MD, PROFESSOR OF
MEDICINE

STATEMENT BEFORE THE UNITED STATES SENATE COMMITTEE ON LABOR
AND HUMAN RESOURCES NOVEMBER 3, 1987

I have concerns regarding the widespread utilisation of aspartame as a
common non-nutritive sweetener (1, 2). Aspartame is a protein-like substance
that is composed of two amino acids (aspartic acid and phenylalanine) and
methanol. I believe the aspartic acid and methanol portions of aspartame are
harmless. However, aspartame usage in high doses can lead to significant
increases in blood phenylalanine. It is this increase in blood phenylalanine
following aspartame consumption that is of concern to me. The basis for my
concerns are three-fold.

1.   Although industry estimates of individual aspartame intake are
approximately 10 mg/kg per day, and the FDAs advisable daily intake (ADI) is
50 mg/kg per day, I believe that a significant portion of the population,
particularly children, will consume upwards of 50 mg/kg per day, or
approximately the ADI. In layman's terms, a dose of 50 mg/kg per day is
approximately 5 servings per 50 lb. body weight per day. While it is true that
most 150 lb. adults will not come close to consuming 15 servings per day, I
believe that children, owing to their reduced body weight, will consume up to
five servings per 50 lb. body weight per day and approximate, or exceed, the
ADI. Indeed 11 year-old studies have shown that when 7-12 year-old children
have free access to a liberal supply of aspartame-sweetened products in their
diet, their average consumption is 50 mg/kg per day and ranges up to 77
mg/kg per day (3). Since aspartame can be found in soft drinks, fruit juices,
breakfast cereals, gelatine, puddings, milk shakes and a whole host of other
products, it is not surprising that a typical 7-12 year-old child may consume
up to five servings per 50 lb. body weight per day. The industry notion that
children consume much less than 10 mg/kg per day must be categorically
rejected, when it is recognised that the consumption of a single 12-oz can of
cola by a 50 lb. 7-year old is already a dosage of 10 mg/kg per day.

2   If it is granted that many children and some adults will consume up
to 50 mg/kg per day of aspartame (i.e., about five servings per 50 lb. body
weight per day) then it should also be recognised that data in the medical
literature indicate that this dosage of aspartame will result in a doubling of
blood phenylalanine in normal individuals and at least a tripling of blood
phenylalanine in subjects who are heterozygous for the phenylketonuric
(PKU) trait (4). The estimate of individuals with PKU heterozygocity (i.e., the
condition associated with slow metabolism of dietary phenylalanine) is
believed to range anywhere from 4-20 million individuals in the United States
(5). Since the normal blood phenylalanine concentration is approximately 50

mM, the blood phenylalanine level will rise to 100-150 mM in individuals who consume approximately 50 mg/kg per day of aspartame (4).

3. My third concern is that there are data in the medical literature which indicate that a tripling of blood phenylalanine to the level of approximately 150-200 mM concentrations may have untoward effects on the human brain. One study suggests that mothers who have a blood phenylalanine concentration of 250 mM bear offspring with a 10 point drop in IQ (5). Another study indicates that when blood phenylalanine is increased to the 250 mM concentration, there is a 10% change in choice reaction time, a test of higher cognitive function in humans (6). More recently, another study has shown that when blood phenylalanine is increased into the range of approximately 150-200 mM, there are quantitative changes in the electroencephalogram in humans (7).

Phenylalanine is a known neuro-toxin, and the food industry added nearly 8,000 tons of aspartame to the food supply in 1986, which amounts to approximately 8,000,000 lb. of phenylalanine. The consumption of aspartame has increased exponentially since its introduction in 1981. The 1986 consumption of aspartame in the United States was equal to nearly 22% of the 1986 consumption of refined sugar (allowing for a 200-fold increase in sweetener potency of aspartame relative to sugar). With this enormous selective infusion of phenylalanine into the food supply, the key questions before the United States Congress and other scientific and medical organisations are whether selective increases in the blood phenylalanine level on the order of 200 mM are to be expected with liberal intake of aspartame, and whether blood phenylalanine increases of this magnitude have untoward effects on the human brain.

Yours very truly, (Signed) William M. Pardridge, MD Professor of Medicine

NEUROLOGIC ASSOCIATES, P.C. HUNTINGTON, NEW YORK; GERALD A. SCHROETER, MD, F.ATA.M

STATEMENT TO SENATOR HOWARD METZENBAUM, SENATE COMMITTEE ON LABOR & HUMAN RESOURCES NOVEMBER 2, 1987

I have been requested to write you regarding my feelings on NutraSweet, both from the aspect of my clinical experience and clinical opinion regarding such. It is my feeling, as a neurologist, that NutraSweet aspartame being a potential neuro-toxin may be a factor involved in neurologic symptoms such as headaches, dizzy spells, fainting episodes or convulsive seizures.

My experience with NutraSweet has been inconclusive, with some of my patients reporting adverse reactions such as headaches, dizzy symptoms and convulsive seizures, possibly triggered by or exacerbated by the usage of this product.

I have read several articles on aspartame, including more recent articles alluding to its complete safety, and "no link" between such and convulsive seizures.

It is my recommendation that this product be further scrutinised and

evaluated objectively, both clinically and experimentally, so that further
conclusions regarding possible adverse reactions can be ascertained. I will
continue to advise my patients with neurologic symptoms to minimise the
usage of this product until I am fully convinced of its safety.
Sincerely, (Signed) Gerald A. Schroeter, MD, F.ATA.M

ATLANTA EYE CLINIC, P.C. ATLANTA, GEORGIA MORGAN B. RAIFORD, MD

STATEMENT TO HONORABLE HOWARD M. METZENBAUM, SENATE LABOR
AND RESOURCES COMMITTEE

Enclosed is the outline regarding the toxicity of aspartame (NutraSweet). This
product has serious side effects from human consumption and should not be
used in the market place.
I had the opportunity, in Atlanta, Ga., to see the effects of methyl alcohol
toxicity in 1952-1953 which resulted in visual damage to the optic nerves and
retina in over 300 cases and the deaths of over 30 persons.
I examined Shannon Roth on July 7, 1986, along with several other patients. I
observed evidence of effects in her eye and the eyes of the other patients that
were comparable to the effects observed in the patients who suffered methyl
alcohol toxicity in 1952-1953.
There was damage in the central fibres, 225,000 of the total 137,000,000 optic
nerve fibres, (resulting in optic nerve atrophy) in her case, which would be
comparable to that observed from patients suffering methyl alcohol toxicity.
The extent of damage to these fibres would explain partial to total blindness.
The macular area of the retina and optic nerve fibres are highly reactive to
the toxicity of methyl alcohol because these fibres and nerve cells require
from four to six times as much oxygen and nutrition as the visual pathways
peripheral nerves. These fibres and nerves (rods and cones) cannot go
without oxygen for more than 90 seconds without some visual loss.
Aspartame is broken down by the upper portion of the digestive tract and
methyl alcohol is then absorbed into the blood vessels of the region which
travels through the bodies[sic] entire vascular system. This toxic effect
causes extensive damage to the human visual pathway, ending with optic
nerve atrophy as well as retinal starvation and visual loss. The intensity of
these toxic effects from the methyl alcohol spin-off causes impaired nerve
damage that varies with intensity as to the amount of methyl alcohol the
individual absorbs.
Each eye will react differently to this kind of involvement. In large doses of
consumption of methyl alcohol the differences in response might not be
noticeable, although, cases of damage in one eye from methyl alcohol have
been reported in the literature.
But in the kind of chronic low dose exposure to methyl alcohol experienced
by Shannon Roth (in NutraSweet consumption) and other NutraSweet
consumers, it is likely that they would experience the impact on the optic
nerve differently in each eye.
The important point is that the damage observed in Shannon Roth's eye was

identical to the damage I observed repeatedly in the eyes of individuals whose eyes have been damaged by methyl alcohol toxicity.

Sincerely,

(Signed) Morgan B. Raiford, MD, B.S., M.Sc Med. Ophthalmology, D.Sc Med Ophthalmology, D.H.L., Am. B. Ophthalmology, Am. College Nuclear Med.

THE RETINA INSTITUTE, MEDICAL AND SURGICAL DISEASES OF THE RETINA AND VITREOUS, ORTON T. AYER, JR., MD, PA

STATEMENT TO SENATE OCTOBER 28, 1987

Mrs. Roth was first examined by me approximately two years ago. At that time she was diagnosed as having optic atrophy and decreased vision in one eye. The other eye was unaffected. All the available medical tests were performed and showed no evidence of an associated neurologic deficit.

Mrs. Roth had been using aspartame in relatively large doses during the time that she experienced visual difficulties. Myself and other doctors question the possibility of aspartame and its breakdown products as being a possible etiologic agent. In reviewing much of the literature available, the presence of methanol even in relatively small quantities can cause neurologic damage in susceptible individuals.

There appears to be considerable controversy concerning NutraSweet. Some of the original investigators who were instrumental in the FDA approval maintain that it is harmless. Certainly, most food additives including MSG and saccharin were thought to be harmless at the time of their introduction. The request for an objective re-evaluation of any substance widely used that could have the potential for harmful effects should be welcomed.

In conclusion, my opinion is that aspartame containing agents can, in selective cases, be neuro-toxic. Mrs. Roth's particular case may indeed be one of many patients who is more sensitive to this additive.

With best regards, (Signed) Orion T. Ayer, Jr., MD. Dictated but not read to expedite delivery.

TESTIMONY GIVEN TO SENATOR HOWARD M. METZENBAUM BY JAMES TURNER, ESQ.

Senator Metzenbaum: Mr. Turner, how do you feel about the way scientific research is conducted, generally, on food additives?

Mr. Turner: The research process on food additives is a weak process. Virtually all the FDA does is summary reviews of studies done by other scientists, and those studies are invariably submitted by industry, and they are presented as a trial brief, an argument why the substance should be allowed in the food supply, not as a balanced analysis of whether or not the substance is safe.

Senator Metzenbaum: Do you believe the key tests that were used in this instance for NutraSweet's approval were scientifically credible?

212

Mr. Turner: At the very least, they are highly controversial, and in particular, one set is of great importance. There are clinical tests that were referred to by the Commissioner, today, that indicated why he could reduce the safety factor from 100 to 40. These clinical tests were submitted by the Bureau of Foods to the Bureau of Drugs for evaluation. The Bureau of Drugs routinely does clinical testing; the Bureau of Foods does not. The Bureau of Drugs reviewed these clinical tests and found them to be unacceptable to be relied on for regulatory purposes. They were returned to the Bureau of Foods with explanations about how to conduct proper clinical tests; the Bureau of Foods decided not to conduct those tests because they were not required for food additives.
(Information supplied follows:)

STATEMENT OF JAMES TURNER

My name is James Turner. I am a consumer interest attorney from Washington, DC I am also the author of 'The Chemical Feast: The Nader Report on the Food and Drug Administration' and 'Making Your Own Baby Food'. I represent the Community Nutrition Institute and the Aspartame Consumer Safety Network in various legal challenges to the approval of aspartame.
My first encounter with aspartame was in 1971 when Dr. John Olney of Washington University in St. Louis, with whom I worked on several food additive issues, informed me that aspartic acid, a component of the then brand new sweetener aspartame, had caused lesions (holes) in the brains of mice he was testing. Both Dr. Olney and I met with representatives of Searle prior to the FDA approval in 1974, but they did not feel that these animal studies raised any health problems.
Dr. Olney and I objected to the 1974 approval based on Dr. Olney's mouse study, another study in which monkeys consuming aspartame had epileptic seizures, and what I alleged to be the poor quality of several other Searle studies. FDA granted our request for a hearing and a years' long battle ensued.
In 1981, the FDA commissioner overturned the decision of a Public Board of Inquiry that more studies be undertaken in the possibility that aspartame might be cancer-causing, and in 1983 FDA extended approval to soft drinks. With the approval of aspartame for soft drinks, followed by reports from several thousand consumers that they had suffered damage from aspartame, a major public controversy ensued.
In June 1983, a month before FDA approved the addition of aspartame to soft drinks, the National Soft Drink Association wrote the FDA Commissioner warning that aspartame breaks down in unexplained ways in warmer climates. The Association said:
"The more persistent, warmer temperatures in some parts of the US will result in more extensive aspartame breakdown. The current petition clearly shows this more complete breakdown of aspartame at temperatures above 30 degrees C. Additionally, the petition indicates inability to account for what

has happened to as much as 30% of the sweetener, despite having analysed for compounds expected in the usual breakdown pathways."

In responding to questions about breakdown product safety, the Commissioner said:

"The agency also concludes that the allegation that all possible reaction products have not been tested for safety is not a tenable issue in terms of regulatory food additive safety evaluation. Such a requirement for the demonstration of safety would necessitate an unlimited amount of experimental data. [48 FR 142 p. 31381]

The Food, Drug and Cosmetic Act requires [21 USC 348(c)(5)] that the FDA consider:

(A) The probable consumption of the additive and of any substance formed in or on food because of the use of the additive;

(B) The cumulative effect of such additive on the diet of man or animals, taking into account any chemically or pharmacologically related substance or substances in such diet. [46 FR 142 p.38287]

Foregoing identification and testing of the 30% of the aspartame that disappears at product temperatures exceeding 80 degrees F may be prudent or defensible. I suspect it is neither. In any case, consumers and legislators need to understand that "safety" as defined in the food and drug law does not apply to 30% of the aspartame additive.

This is just one specific aspect of aspartame's approval that raises doubts about its asserted "safety." The tests that suggested a cancer-causing possibility to the Public Board of Inquiry have not been convincingly rebutted. The test in the 1974 petition in which aspartame caused seizures in monkeys, and which was the subject of a request for a Grand Jury investigation, has yet to be adequately explained.

Raising the acceptable daily intake to make room for soft drinks; overruling official recommendations that marketing be delayed; arguing that withholding aspartame from soft drinks "provides an extra cushion for purposes of Searle's current position" and then relying on the same data to add aspartame to soft drinks; all undermine confidence in the soundness of the science relied on by FDA to approve aspartame.

It will be a long time before the aspartame controversy goes away. But there are currently many other food additives waiting in the wings, new sweeteners, salt and fat substitutes and others which hopefully can be handled in a way that builds rather than undermines public confidence. Setting aside for the moment the specific controversies about aspartame, I hope we can all agree that any actions that can build a general confidence in the food safety decisions of the FDA will help us all.

With this idea in mind I also think there are three general areas that need attention: the scientific testing done before an additive is approved; the manner in which doubts, limitations and possible dangers are communicated to the public through labelling and advertising; and the manner in which an additive, once approved for the market, is monitored for possible dangers.

The following proposals are suggestions made to start a dialogue, not to present an inflexible position. Much of the thinking that has gone into them was done by me when I served as a founding board member of the Food

Safety Council – a group made up of thirty-five food companies interested in food safety. It had a board membership which was fifty percent from the food industry and fifty percent public. I believe that the FSC was an important effort, and much of its work will prove relevant to the emerging food safety debate.

With regard to testing prior to approval, it appears to me that there is one central problem. Individual scientists cannot do their best and most objective work when they are financially beholden to the companies who stand to make hundreds of millions, or even billions of dollars based on the outcome of their research. (One Searle-supported aspartame researcher is reported to have received $1.3 million for his laboratory.)

It is not necessarily a matter of conscious objectivity that poses the problem. Nor is it the problem that enough money will convince a scientist to say that an obviously dangerous additive should be treated as safe. The problem is that in the emerging scientific areas such as brain chemistry, scientific decisions are often a matter of subtle judgements.

It is naive to argue that a subtle observation, on the outcome of which rides over a million dollars of laboratory research money, can be made objectively even by the most strong-willed. Indeed, the whole basis of the American system is the opposite – give a scientist enough money and he will find what you ask him to find – a cure for cancer, an efficient fuel, a spaceship. Entrepreneurial science is sound when you are looking for cures, products, breakthroughs. It does not work well, as currently organised, when you are looking for weaknesses in products, putting brakes on the marketing team, or in other ways limiting money-making potential.

My suggestion is to get the entrepreneurial spirit working for safety. The FDA, or another properly designated agency could certify a series of laboratories as capable of doing the various kinds of testing necessary for product approval – the agency Red Book sets these out now.

Today when a company does the designated test, it hires and pays the laboratories directly. I propose that this relationship be replaced by having the company go to the FDA with its new additive, and the FDA would then assign on a random basis the test required to the previously certified laboratories. The companies would pay the FDA, who would then pay the laboratories.

With regard to information to the public, it is necessary to recognise as Dick Hall, Research Vice President of the McCormick Company said at one of the final meetings of the Food Safety Council, food safety has become personal, subjective and relative. We can no longer, if we ever did, deprive the whole population of some substance just because some small part of the population will be harmed by it.

On the other hand, it is not proper to lead the entire population to believe that an additive will hurt no one merely because it does not hurt everyone. In this situation it is absolutely essential that every single sub-group of the population that might be at risk from an additive be told of that possibility – as much as possible through labelling and advertising warnings. If the full information cannot fit into ads or onto labels then the ads and labels need to tell consumers where to get the detailed information.

Finally, once a substance reaches the market it should not be treated as sacrosanct. It must be recognised that over time a substance that we know harms people will continue to harm people, whereas a substance that we believe might not harm people sometimes turns out to be harmful. My suggestion is that we account for this possibility by creating a rigorous post-marketing surveillance system. This can be accomplished by putting in place a prospective epidemiologic system – by identifying a properly constituted sample of individuals who use the product and following them for five, ten, twenty or more years so that as soon as there is any suggestion of harm to the population it can be discovered.

The proposals I am presenting here are meant to strike a balance between those consumers that want to consume a large number and wide variety of food additives for various purposes, and those (not necessarily different consumers) who may face the possibility of harm from an additive that poses no danger to many other people.

If the standard of food safety is that a substance that harms some people, but not all people, is going to be allowed on the market, then special policies need to be adopted that will protect those at risk.

Thank you for the opportunity to testify here today.

APPENDIX 4

ASPARTAME QUESTIONNAIRE

Name:_____

Age:_____

Sex: (circle one)   M   F

Address:_____

Telephone Numbers

Day: _____   Night:_____

Fax:_____Other: _____

E-Mail:_____

How long have you used
Aspartame?_____

What types of products containing aspartame do you
use?_____

On the average, how much aspartame do you consume
daily?_____

What health symptoms or problems have developed since using
aspartame?_____

Have you consulted a physician about any health problems related to
aspartame?_____

If yes, was your doctor aware of problems associated with
aspartame?_____

Are you insulin dependent or adult-onset
diabetic?_____

Are you
hyperactive?_____

Have you ever consulted a certified
nutritionist?_____

Do you exercise
regularly?_____

Have you tried the aspartame challenge
test?_____

THE ASPARTAME CHALLENGE TEST

Stop using all food and drink containing aspartame for thirty days. Make sure to read all product labels to check for "hidden" aspartame. Not all food products containing aspartame are labelled "sugar-free." Immediately begin recording any health changes. Note all health symptoms that have lessened, disappeared, or remained the same.

Please contact Starr Resources International for additional information about aspartame by mail, e-mail or by visiting our website: http://www.sweetpoison.com.

Please mail your completed questionnaire and any information you have to:

Starr Resources International PO Box 971 McKinney, Texas 75070-0971

References

'Aspartame-Not for the Dieting Pilot?' Aviation Safety Digest, Spring 1989.

'Aspartame Safety Review,' Food Magazine, April/June 1990.

'The Atlas of Endangered Species', Toronto: Belitha Press Unlimited, 1993.

Balch, J.F., and P.A., 'Prescription for Nutritional Healing, Garden City Park: Avery Publishing Group, 1990.

Blaylock, R., 'Excitotoxins: The Taste That Kills,' Santa Fe: Health Press, 1994.

Boehm and Bada, 'Racemization of Aspartic Acid and Phenylalanine at 100* C,' Proceedings of the National Academy of Science, Amino Acid Dating Laboratory, Scripps Institution of Oceanography, University of California at San Diego,Vol. 81, pp. 5236-5266, 1984.

Burton Goldberg Group, The, 'Alternative Medicine - The Definitive Guide,' Tiburon: Future Medicine Publishing, 1993.

Coleman, M.,Jr. 'Peds', 78:985-990, 1971.

'Diet Soda Issue: Do You Know What You're Drinking?' Idea Today, September 1991.

Dow-Edwards, D., 'The Effects of Aspartame on Development (Guinea Pigs)' Project #5 RO1 NS22766-03, Brooklyn: SUNY Health Science Centre, 1989.

Erlichman, J., 'Nutrasweet Alarm Raised 16 Years Ago,' The Guardian, July, 20, 1990.

Erlichman, J., 'Nutrasweet Test Results 'Faked',' The Guardian, July, 20, 1990.

Estes, C.P., 'Women Who Run With the Wolves,' New York: Ballantine, 1992.

Fisher and Rancho, 'Fit to Fly,' Canadian General Aviation News, March 1990.

Fredericks, J., 'Panelists Differ on NutraSweet,' Denton Record Chronicle, Nov. 9, 1991.

Gaines and Bada, 'Aspartame Decomposition and Epimerization in the Diketopiperazine and Dipeptide Products as a Function of pH and Temperature,' Journal of Organic Chemistry, Vol. 53, p. 2757, 1988.

Gold, Fowkes & Dean, 'Aspartame,' Smart Drug News, 1995.

References

Hays, J., studies and case histories of negative effects of aspartame.

Hicks, M., 'Nutrasweet... too good to be true?' General Aviation News, July 89

'High on High,' Plane & Pilot, January 1990.

Jacobs, J., 'Teacher's Health Attacked By Aspartame, She Says,' The North Texas Daily, p. 6, Nov. 13, 1991.

Kolehmainen, J. a Handwork, S., 'Teen Suicide,' 1995.

Lewis, C.C., 'Really Important Stuff My Kids Have Taught Me,' New York: Workman Publishing Co., 1995.

Millstone, E., 'Can Aspartame Be Linked to Brain Tumors?' Science Policy Research Institute, University of Sussex, England, 1996.

Millstone, E., 'Aspartame linked to brain tumors' Science Policy Research Institute, University of Sussex, England, 1996.

Mohler, S., 'Medication and Flying: A Pilots Guide,' Aircraft Owners and Pilots Association.

Monte, W. C., 'Aspartame Methanol and the Public Health,' Journal of Applied Nutrition, 36 (1) 42-53, 1984.

Moser, R. H., 'Serendipity Can Be Nudged,' The Pharos, Vol. 54, No.3, p31, Summer 1991

Moskal, P., 'Study of George Leighton/effects on pilots while flying,' 1990.

Mullarkey, B., 'Aspartame Facts & Figures,' Nutrimice, Spring 1989.

Mullarkey, B., 'Bittersweet Aspartame, A Diet Delusion,' Nutrvoice, 1992.

'Nutrasweet-Health and Safety Concerns,' Senate Hearing No. 100-567, Nov. 3, 1987.

'Nutrition Action,' Washington, DC: Centre for Science in the Public Interest,June 1991.

Olney,J.W and Ho, O., 'Brain Damage in Infant Mice Following Oral Intake of Glutamate, Aspartate, or Cysteine,' Nature' 227: 609-610, 1970.

'Physicians Feeling Squeezed Between a Rock and a Hard Place? Ineffective Treatment for Chronic/Recurrent Illness, American Academy of Environmental Medicine, April 17-22, 1997.

'Pilot Safety Fears,' Food Magazine, April/June 1990.

'The Quiet Mind,' Sayings of White Eagle, Oxford: University Printing House, 1972.

Remington, D. and Higa, B., 'The Bitter Truth About Artificial Sweeteners,' Provo: Vitality House International, 1987.

Roberts, H.J., 'Aspartame and Hyperthyroidism: A Presidential Affliction Reconsidered,' Townsend Letter for Doctors and Patients, May 1997.

Roberts, H.J., 'Aspartame (NutraSweet) – Is It Safe?,' Philadelphia: The Charles Press, 1990.

Roberts, H.J., 'Clinical Research,' Vol. 36, No. 3, 489A, 1988.

Roberts, H.J., 'Sweet'ner Dearest: Bittersweet Vignettes,' West Palm Beach: The Sunshine Sentinel Press, 1992.

Rohn, E.J. and Reynolds, R.L., The Seasons of Life, Dallas: Total Impact, 1981.

'Safety of Amino Aads,' Life Sciences Research Office, FASEB, FDA Contract No. 223-88-2124, Task Order No. 8.

Smith, L., 'Greed vs. Health – Which Will Win? The Facts,' 1991.

Stoddard, M., 'Deadly Deception The Story of Aspartame,' Dallas: Aspartame Consumer Safety Network, 1996.

Stoddard, M. & Leighton, G., 'Aspartame and Flying,' Extraordinary Science, Spring 1995.

Stoddard, M. & Leighton, G., 'Aspartame alid Flying, The Incredible Untold Story,' 1997.

'This Could Save Your Life,' Pacific Flyer, November 1988.

Toft, P., letter written to athletic trainers about aspartame's effects on athletes, 1991.

Tsang, Clarke and Parrish, 'Determination of Aspartame and Its Breakdown Products in Soft Drinks by Reverse – Phase Chromatography with W Detection,' Journal of Agricultural and Food Chemistry, Vol.33, p. 734-748; New Orleans: Sugar Processing Research, Inc., and the US Department of Agriculture, ARS, Southern Regional Research Centre, 1985.

US Court of Appeals for the District of Colombia Circuit, No. 84-1153, Sept. 24, 1985.

Wall, S. and Arden, H., 'Wisdomkeepers,' Hillsboro: Beyond Words Publishing, 1995.

Wurtman and Walker, 'Dietary Phenylalanine and Brain Function,' First International Meeting on Dietary Phenylalanine and Brain Function, Washington, DC: May 8-10, 1987.